Annals of Mathematics Studies
Number 198

The Plaid Model

Richard Evan Schwartz

PRINCETON UNIVERSITY PRESS

PRINCETON AND OXFORD

2019

Published by Princeton University Press,
41 William Street, Princeton, New Jersey 08540

In the United Kingdom: Princeton University Press,
3 Market Place, Woodstock, Oxfordshire OX20 1SY

LCCN: 2018942719

ISBN 978-0-691-181370

ISBN (pbk.) 978-0-691-18138-7

British Library Cataloging-in-Publication Data is available

Editorial: Vickie Kearn, Lauren Bucca, and Susannah Shoemaker
Production Editorial: Nathan Carr
Production: Jacquie Poirier
Publicity: Alyssa Sanford and Kathryn Stevens

This book has been composed in LATEX

The publisher would like to acknowledge the author of this volume for providing the camera-ready copy from which this book was printed.

Printed on acid-free paper. ∞

press.princeton.edu

Printed in the United States of America

10 9 8 7 6 5 4 3 2 1

Contents

Preface

The purpose of this monograph is to study a construction, based on elementary geometry and number theory, which produces for each rational parameter (satisfying some parity conditions) a cube filled with polyhedral surfaces. When the surfaces are sliced in one direction, the resulting curves encode all the essential information about the so-called special outer billiards orbits with respect to kites. When the surfaces are sliced in two other directions, they encode all the essential information in a 1-parameter family of the Truchet tile systems defined in [**H**].

I call the construction the plaid model. The reason for the name is that plaid shirts involve a network of horizontal and vertical lines, but the underlying weave in this shirt is slanting. The definition of the plaid model involves these kinds of lines. Also, a very plaid-like pattern of lines appears when one does calculations with the model. See Figure 3.3.

The plaid model grew out of my work in [**S1**], where I gave an affirmative answer to the Moser-Neumann question about outer billiards: *Does there exist an outer billiards system with an unbounded orbit?* The main result of [**S1**] is that outer billiards has unbounded orbits relative to any irrational kite – a bilaterally symmetric convex quadrilateral which is not affinely equivalent to a lattice polygon. This monograph is in some ways a sequel to [**S1**], though it can be read independently from [**S1**].

At least to me, the plaid model has a physical feel, with properties that seem like conservation laws, interacting particles, spacetime diagrams, and even an exclusion principle. (I don't claim that the plaid model actually models something in the physical world.) The plaid model also has an overtly hierarchical structure, which causes it to exhibit properties such as self-similarity and scaling limits. Finally, it has an interpretation in terms of a higher dimensional polytope exchange transformation.

This monograph establishes some of the basic properties of the plaid model: the connection to outer billiards and to Truchet tilings, the connection to polytope exchange transformations, and some results about the size and distribution of the polygons in the slices of the model. I hope that this monograph brings out the beauty, depth, and surprise of the plaid model and also suggests topics for further study.

A novel feature of the monograph is that it comes with a companion computer program which illustrates all the main results and constructions. At the end of the introduction I give instructions for downloading the program, and throughout the monograph I make comments on how to use the program to see the relevant points discussed in the text.

I thank the National Science Foundation for their continued support, and also the Simons Foundation for a Simons Sabbatical Fellowship during which I worked on this monograph. I also thank Peter Doyle, Pat Hooper, John Smillie, Sergei Tabachnikov, and Ren Yi for a number of conversations related to the plaid model.

Introduction

The plaid model grew out of my attempt in [**S1**] to understand outer billiards on kites. A *kite* is a convex quadrilateral having a line of symmetry that is also a diagonal. In particular, let K_A be the kite with vertices

$$(-1,0), \quad (0,1), \quad (0,-1), \quad (A,0). \tag{1}$$

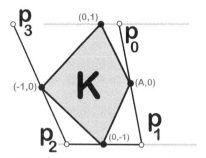

Figure 0.1: Outer billiards on the kite K_A.

Figure 0.1 shows outer billiards on K_A for $A = 4/9$. Given $p_0 \in \mathbf{R}^2 - K_A$, we define a map $p_0 \to p_1$ by the rule that the line segment $\overline{p_0 p_1}$ is tangent to K_A at its midpoint, and K_A is on the right-hand side as one walks along the segment from p_0 to p_1. We then consider the orbit $p_0 \to p_1 \to p_2$.... See [**S1**] for an extensive discussion of outer billiards and a long bibliography.

We call an outer billiards orbit on K_A *special* if it lies in the union

$$\mathbf{R} \times \{\pm 1, \pm 3, \pm 5, ...\} \tag{2}$$

of odd-integer horizontal lines. The orbit shown in Figure 0.1 is special. In [**S1**] I proved the following result.

Theorem 0.1 *When A is irrational, outer billiards on K_A has an unbounded special orbit.*

Theorem 0.1 is an affirmative answer to the *Moser-Neumann problem*, from 1960, which asks whether an outer billiards system can have an unbounded orbit. The orbits in Theorem 0.1 are quite complicated. They return infinitely often to every neighborhood of the vertices of K_A. I called such orbits *erratic*.

The key step in understanding the special orbits on K_A is to associate an embedded lattice polygonal path to each special orbit. This path encodes the symbolic dynamics associated to the second return map to the union $\boldsymbol{R} \times \{-1, 1\}$ of lines. These lines are partially shown in Figure 0.1. When $A = p/q$ is rational, it is possible to consider the union of all these lattice paths at once. I call this union the *arithmetic graph* and denote it by Γ_A. When pq is even, every component of Γ_A is an embedded lattice polygon. Part 3 of the monograph has a detailed description.

One of the key results in [**S1**] is the *Hexagrid Theorem*. This result gives large-scale structural information about Γ_A. Basically, it says that Γ_A must intersect certain lines in certain places, and must avoid certain lines in certain places. Some years later I discovered that the Hexagrid Theorem is just the first in a series of results which allowed this large-scale structure to extend down to increasingly fine scales. When all these results are assembled into one package, the result is the plaid model.

We will formally define the plaid model in the next chapter. The plaid model is a rule for assigning a square tiling of the plane to each parameter $A = p/q \in (0, 1)$ with pq even. We call such parameters *even rational*. There is a similar construction when pq is odd, but the details are sufficiently different that we do not treat it here. Here we give a rough feel for the plaid model. Based on the parameter A we assign even integers to the lines of the usual infinite grid of integer-spaced vertical and horizontal lines. We call these integers *capacities*. At the same time, we define a second grid of slanting lines and we assign odd integers to these lines. We call these odd integers *masses*. We then place a *light point* at every intersection of the form $\sigma \cap \tau$ where

- σ is a slanting line.

- τ is a horizontal or vertical line.

- The mass of σ has the same sign as the capacity of τ and smaller absolute value.

Figure 0.2 illustrates the rule on a made-up example. The pictures in the next chapter show the real rules.

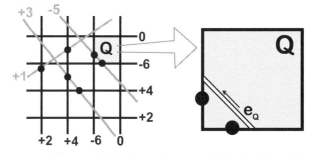

Figure 0.2: A caricature of the plaid model.

The horizontal and vertical lines divide the plane into unit integer squares. It turns out that each such square Q has either 0 or 2 edges containing an odd number of light points. In the former case we associate the empty set to Q. In the latter case, we associate to Q a directed edge e_Q which joins the centers of the two sides having an odd number of light points. (We will explain later how the edge direction is determined.) Figure 0.2 shows a made-up example of the assignment $Q \rightarrow e_Q$.

The edges fit together to form an infinite family PL_A of polygons in the plane which we call the *plaid polygons*. Again $A = p/q \in (0,1)$ and pq is even. The lines of capacity 0 divide the plane into larger squares of side length $p + q$, which we call *blocks*. No polygon crosses the boundaries of these blocks. Figure 0.3 shows two of the blocks associated to $PL_{4/9}$. We do not show the orientations of the edges but they are consistently oriented around each polygon, one way or the other.

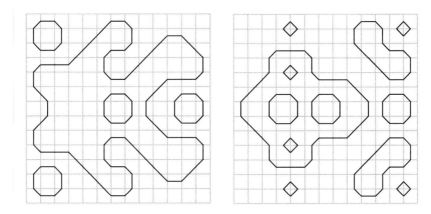

Figure 0.3: Two blocks of $PL_{4/9}$.

Here is a concrete connection between the plaid model and outer billiards.

Theorem 0.2 (Projection) *Let $A \in (0,1)$ be an even rational parameter. Modulo the vertical translations which preserve PL_A, there is a bijection between the polygons in PL_A that lie in the right half-plane and the special outer billiards orbits relative to the kite K_A. Moreover, the plaid polygon π may be (monotonically) parameterized as $\pi = \{(x_t, y_t)| \ t \in [0, N]\}$ in such a way that the point $2x_k$ lies within 3 units of the kth point of*

$$S_\pi = O_\pi \cap (\mathbf{R}_+ \times \{-1, 1\})$$

for all $k \in \{1, ..., N\}$. Here O_π is the special orbit associated to π and N is the number of points S_π.

In other words, if you put your finger on one of the polygons π of PL_A (that lies to the right of the Y-axis) and trace around it at the correct speed, the horizontal motion of your hand will track the first return map of the corresponding special outer billiards orbit up to a factor of 2 and an error of at most 3 units. Note, however, that π typically does not have N vertices and x_k need not be a vertex of π.

The Projection Theorem is a consequence of the Quasi-Isomorphism Theorem below, which gives a more precise result about the connection between PL_A and Γ_A. The Quasi-Isomorphism Theorem is the multiscale extension of the Hexagrid Theorem from [**S1**].

The plaid model has a three-dimensional interpretation which reveals connections to Pat Hooper's Truchet tile system [**H**]. When we forget the orientations on the polygons, it turns out that there are $p + q$ distinct blocks modulo translation symmetry of the tiling. (When we remember the orientations there are twice as many.) We will take $p + q$ such blocks, one representative from each translation equivalence class, and stack them on top of each other in a special order. We will then canonically interpolate between the polygons at consecutive heights in the stack to form polyhedral surfaces. The result is a cubical array of $(p + q)^3$ unit integer cubes that is filled with pairwise disjoint embedded polyhedral surfaces.

By construction, the slices of the 3D polyhedral surfaces at integer heights in the XY direction are the plaid model polygons discussed above. When the surfaces in the plaid model are sliced in the other coordinate directions, namely the XZ and YZ directions, what emerges (at least for some slices) is a pattern of curves that is combinatorially isomorphic to the curves produced by Pat Hooper's system. Figure 0.4 gives an example. The plaid parameter is 4/9 and the Truchet parameter is $\alpha = \beta = 3/8$. Theorem 7.2, the Truchet Comparison Theorem, establishes a combinatorial isomorphism like this for all even rational parameters. Thus, the plaid model is a kind of marriage between outer billiards on kites and the Truchet tile system.

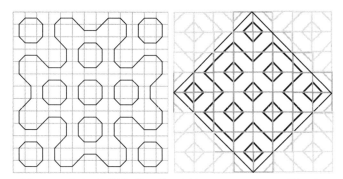

Figure 0.4: A YZ slice for 4/9 compared to a Truchet tile.

Another curious connection between the plaid model and the Truchet tile system is that the left edge of the figure on the left side of Figure 0.3, which shows the union of plaid polygons in a certain block for the parameter 4/9, exactly matches the left edge of the left side of Figure 0.4, which shows a special YZ slice for the same parameter 4/9. This seems to happen for every parameter. We will give several plausible explanations in §23.8, but we will stop short of giving a proof.

The monograph has 5 parts. The rest of this introduction is a detailed description of these parts.

0.1 PART 1: THE PLAID MODEL AND ITS PROPERTIES

In Part 1, I will define the plaid model and study its properties. One technical claim, Theorem 1.4, will not be apparent from any of the definitions. This result basically says that each unit integer square has 0 or 2 sides containing exactly one light point. I will assume Theorem 1.4 in Part 1 and will deduce it in Part 2 as an immediate corollary of Theorem 8.2.

After studying the basic properties of the model, I will explain how one can use the hierarchical nature to get information about the large-scale structure of the tilings in an algorithmic way. In particular, I will give a heuristic explanation of why the model exhibits coarse self-similarity and rescaling phenomena.

After giving the basic definitions, I will explain how to assemble the two-dimensional blocks into embedded polyhedral surfaces. Finally, I will establish the connection between the XZ and YZ slices of these surfaces and the Truchet tilings. Again, the main result is the Truchet Comparison Theorem from §7.3.

Part 1 is rather long and involved, but most of the material is not needed for Parts 2-4. The (dis)interested reader can skip to Part 2 after reading §1, §2.2, §2.3, §4.2, §4.3, §4.4, and §5.2.

0.2 PART 2: THE PLAID PET

Let Π denote the set of unit integer squares. For each parameter $A = p/q$, the union of plaid polygons PL_A defines a dynamical system on Π. We simply follow the directed edge in each tile and move to the tile into which the edge points. When the tile is empty, we do not move at all. We call this dynamical system the PL_A-*dynamics*. This system is similar to the *curve-following dynamics* defined in [**H**].

In Part 2, I will connect this dynamical system to higher dimensional polytope exchange transformations. The parity claim from Part 1 will follow from this. Let X be a flat torus. A *polytope exchange transformation* (or *PET*) on X is given by a partition of X into polytopes

$$X = \bigcup A_i = \bigcup B_i, \tag{3}$$

so that there are translations T_i such that $T_i(A_i) = B_i$ for all i. Such a system gives rise to a global and almost everywhere defined map $T : X \to X$ defined so that $T|_{A_i} = T_i$. This map is not defined on the boundaries of the polytopes of the partition. However, it is an invertible piecewise defined translation.

Now we describe what we mean by a *fibered integral affine PET*. Let

$$\widehat{X} = \boldsymbol{R}^3 \times (0,1).$$

We will work with a quotient of the form $X = \widehat{X}/\Lambda$, where Λ is a discrete group of affine transformations acting on $\widehat{X}$. The quotient X is topologically the product of a 3-torus and $(0, 1)$. The group Λ preserves each slice $\boldsymbol{R}^3 \times \{P\}$ and acts there as a group of translations. The quotient $X_P = (\boldsymbol{R}^3 \times \{P\})/\Lambda$ is a flat torus whose isometry type depends on P.

In a fibered integral affine PET, we have the same partitions as above, except that each map T_i is a locally affine map, and we have the following additional features:

- The linear part of T_i is independent of i.

- T_i preserves each slice X_P.

- The restriction of each T_i to X_P is a translation.

- All vertices of all lifts of all polytopes in the partitions have integer coordinates.

Theorem 0.3 (Plaid Master Picture) *There is a 4-dimensional fibered integral affine PET X with the following property: When A is even rational and $P = 2A/(1 + A)$, there is a locally affine map $\Phi_A : \Pi \to X_P$ which conjugates the PL_A-dynamics on Π to the PET dynamics on X_P.*

Remarks: (i) Since Π is a discrete set of points, we have to say what we mean by a *locally affine map* from Π into X_P. We mean a restriction of a planar affine map to Π.
(ii) The Plaid Master Picture Theorem says that the PL_A-dynamics encodes the symbolic dynamics associated to a certain 3-dimensional PET, X_P. Here $P = 2A/(1 + A)$. It might seem a bit funny to use the parameter P instead of A but this change of coordinates turns out to be useful and natural.
(iii) The Plaid Master Picture Theorem also says that these individual slices $\{X_P\}$ fit together into a 4-dimensional fibered integral affine PET. This shows a kind of coherence between the plaid model at one parameter and the plaid model at a different parameter, even though the plaid model polygons themselves vary wildly from parameter to parameter.

0.3 PART 3: THE GRAPH PET

In Part 3 we do for the arithmetic graph what we did for the plaid polygons in Part 2. When $A = p/q$ is rational, the arithmetic graph Γ_A defines a dynamical system on $\boldsymbol{Z}^2$. We just move from vertex to vertex according to the oriented polygons. We call this system the Γ_A-*dynamics*. Here is our main result.

Theorem 0.4 (Graph Master Picture) *There is a 4-dimensional fibered integral affine PET Y with the following property. When $A = (0, 1)$ is rational, there is a locally affine map $\Psi'_A : \boldsymbol{Z}^2 \to Y_A$ which conjugates the Γ_A-dynamics on $\boldsymbol{Z}^2$ to the PET dynamics on Y_A.*

Remarks: (i) In this result, A need not be *even* rational.
(ii) We will see that there is a change of coordinates that conjugates Ψ'_A to a map Ψ_A with a nicer formula. That is the reason for our notation.

I will explain Theorem 0.4 in two ways. First, I will deduce Theorem 0.4 from [**S1**, Master Picture Theorem], which is presented here (with a few cosmetic changes) as Theorem 13.2.

Second, I will follow [**S2**] and my unpublished preprint [**S3**], and prove a very general version of Theorem 0.4, namely Theorem 16.9, which works for any polygon without parallel sides. (The no-parallel-sides condition is not really essential, but it makes the argument easier.) Since Theorem 0.4 already follows from Theorem 13.2, I will not give a formal proof that the PET produced by Theorem 16.9, in the case of special orbits on kites, is identical to the PET from Theorem 0.4. However, in §16.7 I will explain precisely how to match up the two PETs, and I will explain how to see a computer demonstration of the match-up.

I'd like to mention that I had many helpful and interesting discussions with John Smillie about this general topic, and he has a different way to view these kinds of compactifications in terms of Dehn-invariant like constructions involving the tensor product.

0.4 PART 4: PLAID-GRAPH CORRESPONDENCE

In this part of the monograph we relate the plaid model to outer billiards. Our next result says that the combinatorially defined plaid model gives a uniformly accurate model of the dynamically defined outer billiards special orbits. When the time comes, we will deduce the Projection Theorem above as a consequence.

Theorem 0.5 (Quasi-Isomorphism) *For each even rational parameter A, there exists an affine transformation $T_A : \mathbf{R}^2 \to \mathbf{R}^2$, and a bijection between the components of $T_A(\Gamma_A)$ and the components of PL_A with the following property. If $\gamma \in T_A(\Gamma_A)$ and $\pi \in PL_A$ are corresponding components then there is a homeomorphism $h : \gamma \to \pi$ which moves points by no more than 2 units.*

Figure 0.5 below shows the Quasi-Isomorphism Theorem in action for the parameter $A = 3/8$. The grey polygons are the components of $T_A(\Gamma_A)$ in $[0, 11]^2$ and the black polygons are components of PL_A in $[0, 11]^2$.

One thing that is surprising about the Quasi-Isomorphism Theorem is that the grid Π of half-integer points associated to the plaid model sits in a funny way with respect to the grid $G_A = T_A(\mathbf{Z}^2)$ which contains the vertices of $T_A(\Gamma_A)$. The grid G_A has co-area $1 + A$ whereas Π has co-area 1. When A is irrational, these grids are incommensurable. The homeomorphism from the Quasi-Isomorphism Theorem does not map vertices to vertices but nonetheless the two sets of polygons are close to each other.

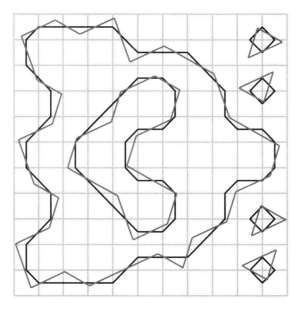

Figure 0.5: The Quasi-Isomorphism Theorem in action.

The Quasi-Isomorphism Theorem is a consequence of an orbit equivalence between our two PETs. Call a subset $Z \subset X$ *dynamically large* if every nontrivial orbit on X intersects Z.

Theorem 0.6 (Orbit Equivalence) *There is a dynamically large subset $Z \subset X$ and a map $\Omega : Z \to Y$ with the following property. For any $\zeta \in Z$ with a well-defined orbit the following three statements hold:*

1. *There is some $k = k(\zeta) \in \{1, 2\}$ such that*
$$\Omega \circ F_X^k(\zeta) = F_Y \circ \Omega(\zeta). \tag{4}$$

2. *If $F_X(\zeta) \in Z$ then there is some $\ell = \ell(\zeta) \in \{0, 1\}$ such that*
$$\Omega \circ F_X(\zeta) = F_Y^\ell \circ \Omega(\zeta). \tag{5}$$

3. *ζ is a fixed point of F_X if and only if $\Omega(\zeta)$ is a fixed point of F_Y.*

The set Z is a union of 2 open convex integral prism quotients and the restriction of Ω to each one is an integral projective transformation that maps the slice Z_P into the slice Y_A, where $P = 2A/(1 + A)$. Finally, Ω is at most 2-to-1 and $\Omega(Z)$ is open dense in Y and contains all the well-defined F_Y-orbits.

A *prism quotient* is the quotient of a set of the form $H \times \mathbf{R} \times [0, 1]$ under the map $(x, y, z, P) \to (x, y, z + 2, P)$, where H is a convex polygon. The prism quotient is *convex integral* if H has integer vertices. The Orbit Equivalence Theorem is saying roughly that the graph PET is a *renormalization* of the plaid PET. We will explain this point of view in §18.8.

0.5 PART 5: THE DISTRIBUTION OF ORBITS

Part 5 concerns the distribution and geometry of orbits in the plaid model.

We will explain how to assign a polygonal path π to an orbit O of the plaid PET which lies in a slice X_A where A is not even rational. The path π is unique up to translation. We call π the *plaid path* associated to O. One could view π as the geometric (i.e., Hausdorff topology) limit of plaid polygons associated to a sequence of orbits $\{O_n\}$ in even rational slices.

We say that a path π in the plane is *thin* if it is contained in an infinite strip. Otherwise we say that π is *fat*. For instance, a parabola is fat. The bulk of Part 5 is devoted to proving the following result.

Theorem 0.7 *For every irrational parameter A, the slice X_A of the plaid PET has an infinite orbit whose associated plaid path π is fat. More precisely, the projection of π onto the X-axis is the set $[1/2, \infty)$ and the union of vertices of π having first coordinate $1/2$ contains a large-scale Cantor set.*

By a *large-scale Cantor set*, we mean a set of the form

$$\{1/2\} \times \bigcup_{\beta} \left(\sum_j \beta_j Y_j \right). \tag{6}$$

The union takes place over all finite binary sequences β, the sequence $\{Y_j\}$ is unbounded, and the ordering on the points in the union coincides with the lexicographic ordering $(0, 1, 10, 11, 100...)$ on the finite binary strings.

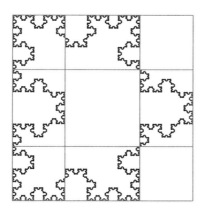

Figure 0.6: Cantor-set like properties of the left edge.

Figure 0.6 shows the picture of a particular plaid polygon for the parameter $169/408$. The square here has side length 577. The extra lines (making a tic-tac-toe pattern) have significance we will explain in the proof of Theorem 3.1. One can see a kind of finite version of a large-scale Cantor set going up along the left edge of the figure. The curve π from Theorem 0.7 is a geometric limit of these finite curves taken with respect to a continued-fraction-like approximating sequence of rational numbers.

Theorem 0.7 combines with the Projection Theorem to give another proof of Theorem 0.1. At the same time, one can extract Theorem 0.7 from the main results in [**S1**] and from the Quasi-Isomorphism Theorem. See §21.6. The ideas underlying our proof here are similar to what we do in [**S1**]. However, by going through the details here we can give a proof that is independent from [**S1**].

We will combine the proof of Theorem 0.7 with some results we prove in Part 1 to get some additional information about the distribution of orbits that goes beyond what we could do in [**S1**]. Say that a polygon is *N-fat* if it is not contained in a strip of width N.

Theorem 0.8 *Let $\{p_k/q_k\} \subset (0,1)$ be any sequence of even rational numbers with a limit that is neither rational nor quadratic irrational. Let $\{B_k\}$ be any sequence of associated blocks. Let N be any fixed integer. Then the number of N-fat plaid polygons in B_k is greater than N provided that k is sufficiently large.*

Remark: The restriction that the limit not be quadratic irrational is probably just an artifact of the proof rather than a necessary hypothesis.

The Projection Theorem lets us translate Theorem 0.8 into the language of outer billiards. Say that the *essential diameter* of a special outer billiards orbit on K_A is the diameter of its intersection with the set

$$[0, \infty) \times \{-1, 1\}.$$

A special orbit of diameter D essentially fills out an octagon-shaped "annulus" of width D.

Corollary 0.9 *Let $\{p_k/q_k\} \subset (0,1)$ be any sequence of even rational numbers with a limit that is neither rational nor quadratic irrational. Let $\omega_k = p_k + q_k$. Let $\{I_k\}$ be any sequence of intervals of the form $[n\omega_k, n\omega_k + \omega_k]$. Let N be any fixed integer. Then there are more than N distinct orbits in the interval $I_k \times \{1\}$ which have essential diameter at least N provided that k is sufficiently large.*

0.6 COMPANION PROGRAM

The monograph comes with several companion computer programs which illustrate most of the results. You can download these from the following location.

https://press.princeton.edu/titles/13339.html

At least in the short run, the same programs can also be found at the following location.

http://www.math.brown.edu/~res/Java/PLAID.tar

Once you download the tarred directory, you untar it. The unpacked files live in a directory called *Plaid*. This directory has several sub-directories and a *README* file. The *README* file has further instructions. One of the programs can also be run directly on the web. At least in the short run, the web version can also be accessed from the following location.

http://www.math.brown.edu/~res/Javascript/Plaid/Main.html

This web program gives a quick view of the 3D plaid model and the plaid surfaces.

I discovered all the results in this monograph using the program, and I have extensively checked my proofs against the output of the program. While this monograph mostly stands on its own, the reader will get much more out of it by using the program while reading. I would say that the program relates to the material here the way a cooked meal relates to a recipe. Throughout the text, I have indicated computer tie-ins which give instructions for operating the computer program so that it illustrates the relevant phenomena. I consider these computer tie-ins to be a vital component of the monograph.

Part 1. The Plaid Model

Chapter One

Definition of the Plaid Model

1.1 CHAPTER OVERVIEW

The goal of this chapter is to define the plaid model. The input to the plaid model is a rational $A = p/q \in (0,1)$ with pq even. The output is the union PL_A of plaid polygons. Our pictures show the case $A = 2/5$ in detail.

In §1.2, we define some auxiliary quantities associated to the parameter. In §1.3 we define 6 families of parallel lines. The first two families are the horizontal and vertical lines comprising the boundary of the integer square grid, and we call these *grid lines*. The remaining 4 families we call *slanting lines*. In §1.4, we explain the assignment of masses and capacities to the lines. In §1.5 we define the concept of a *light point*. The light points are certain intersections between slanting lines and grid lines. We will actually give two different definitions and prove that they are equivalent. The two definitions highlight different features of the model. In §1.6 we assign unit vectors to the light particles. Finally, in §1.7 we put everything together and give the definition of PL_A.

1.2 BASIC QUANTITIES AND NOTATION

We fix an even rational p/q as above. Here are the main auxiliary quantities associated to these parameters.

$$\omega = p + q, \qquad P = \frac{2p}{\omega}, \qquad Q = \frac{2q}{\omega}. \qquad (1.1)$$

Note that $P + Q = 2$ and $P/Q = p/q$.

We define $\widehat{\tau} \in (0, \omega)$ and $\tau \in (0, \omega/2)$ to be such that

$$2p\widehat{\tau} \equiv 1 \mod \omega, \qquad \tau = \min(\widehat{\tau}, \omega - \widehat{\tau}). \qquad (1.2)$$

Congruence Notation: Given some integer N and some integer a, we define $(a)_{2N}$ to be the representative of $a \mod 2N$ in $(-N, N)$. If $a \equiv N \mod 2N$ we define $a_{2N} = *N$, a symbol which denotes the set $\{N, -N\}$.

1.3 SIX FAMILIES OF LINES

We consider 6 infinite families of lines.

- $\mathcal{H}$ consists of horizontal lines having integer y-coordinate.

- $\mathcal{V}$ consists of vertical lines having integer x-coordinate.

- $\mathcal{P}_\pm$ is the set of lines of slope $\pm P$ having integer Y-intercept.

- $\mathcal{Q}_\pm$ is the set of lines of slope $\pm Q$ having integer Y-intercept.

We call the lines in $\mathcal{H}$ and $\mathcal{V}$ the *grid lines*, and we call the other lines the *slanting lines*. Until §1.5 we only use the 4 families $\mathcal{V}$, $\mathcal{H}$, $\mathcal{P}_-$ and $\mathcal{Q}_-$. We set $\mathcal{P} = \mathcal{P}_-$ and $\mathcal{Q} = \mathcal{Q}_-$ in order to simplify the notation. Figure 1.1 shows these lines inside $[0,7]^2$ for $p/q = 2/5$. In this case, $P = 4/7$ and $Q = 10/7$.

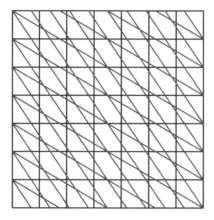

Figure 1.1: The 4 line families for $p/q = 2/5$.

Fixing the parameter p/q, we define a *block* to be the image of a unit integer square under the dilation $(x,y) \to \omega(x,y)$. For instance, $[0,\omega]^2$ is a block. The pattern of lines we have defined is precisely the same in each block.

Say that a *horizontal block segment* is the intersection of a block with a horizontal grid line. Note that there are two slanting lines meeting at the endpoints and the midpoint of each horizontal block segment. We call these points *double points*.

Computer Tie-In: Open up the main program. When the main program is running, open up the *planar window*. On the *plaid model* control panel, turn off all the features except *grid*, *(+) slant lines* and *(-)slant lines*. This will show you the 6 families of lines. You can choose other parameters either by pressing the *random* button at the top right or else by clicking on the blue box at the top right and then entering a fraction with the keyboard. (If you open the *document* window and click on the question box in the top right corner, you can learn more about entering parameters.) Click the left/right mouse buttons or (z,c) arrow keys to zoom into the picture in the *planar window*.

1.4 CAPACITY, MASS, AND SIGN

We associate *signs*, *capacities*, and *masses* to our lines as follows.

- The *signed capacity* of the grid line $y = y_0$ is $(4py_0)_{2\omega}$.

- The *signed capacity* of the grid line $x = x_0$ is $(4px_0)_{2\omega}$.

- The *signed mass* of a slanting line through $(0, y)$ is $(2py + \omega)_{2\omega}$.

Sometimes we will want to consider the signs or the absolute values of the quantities just defined. We define the *capacity* of a grid line to be the absolute value of its signed capacity. We define the *sign* of a grid line to be the sign of its capacity. Likewise we define the quantities *mass* and *sign* for the slanting lines. When the capacity is 0, the sign is indeterminate. Note that the capacity (being even) is never $\pm\omega$, so it is always well defined.

Figure 1.2 shows the assignment of these quantities to the lines in the square $[0, 7]^2$ (and also to a few additional lines). The black labels are the signed capacities and the grey labels are the signed masses. For instance, the slanting line of slope $-4/7$ through the point $(0, 5)$ has mass -1 and the slanting line of slope $-10/7$ through $(0, 8)$ has mass -3.

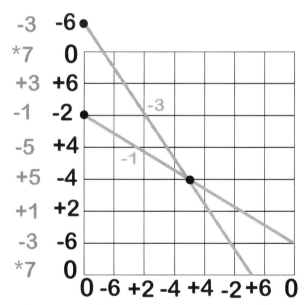

Figure 1.2: The mass and capacity labels for $p/q = 2/5$.

In our example $\hat{\tau} = 2$. Note that the lines $y = 0, 2, 4, 6$ and $x = 0, 2, 4, 6$ have signed capacity $0, 2, 4, 6$ respectively, and the slanting lines through $(0, 0)$, $(0, 2)$, $(0, 4)$, $(0, 6)$ have signed mass $*(0 - 7), 1, (2 - 7), 3$ respectively. The following lemma generalizes this observation.

Lemma 1.1 *Let $k \in \{0, ..., (\omega - 1)/2\}$. Let $\zeta \equiv \pm k\widehat{\tau}$ mod ω. The grid lines $x = \zeta$ and $y = \zeta$ have capacity $2k$. The slanting line through $(0, \zeta)$ has mass k when k is odd and mass $\omega - k$ when k is even. Moreover, the signs of the relevant grid lines match the signs of the relevant slanting lines when k is odd and mismatch when k is even and positive.*

Proof: We will deal with the signs at the end. We consider horizontal grid lines first. The assignments of mass and capacity are invariant under the operations $\zeta \to \zeta + \omega$ and $\zeta \to -\zeta$.

So we just have to deal with the grid line $y = k\widehat{\tau}$. Note that $2p\widehat{\tau} \equiv 1 + \omega$ mod 2ω because $2p\widehat{\tau}$ is even. This congruence gives us

$$4py = (2p\widehat{\tau})(2k) \equiv 2k \text{ mod } 2\omega. \tag{1.3}$$

Hence $(4py)_{2\omega} = 2k$. The proof for the vertical grid lines is the same.

Thanks to the same operations as discussed in the vertical case, we just have to deal with the slanting lines through $(0, k\widehat{\tau})$. Setting the Y-intercept to be $y = k\widehat{\tau}$, we know that the mass of our line is $(2py + \omega)_{2\omega}$. We compute

$$2py + \omega = (2p\widehat{\tau})k + \omega \equiv k + (k + 1)\omega \text{ mod } 2\omega. \tag{1.4}$$

When k is odd, $k + 1$ is even and the quantity in Equation 1.4 is congruent to k mod 2ω. When k is even, $k + 1$ is odd, the the quantity in Equation 1.4 is congruent to $\omega - k$ mod 2ω.

The statement about signs excludes the case $k = 0$, so we take $k > 0$ and otherwise as above. The operation $\zeta \to \zeta + \omega$ preserves the signs of all relevant lines and the operation $\zeta \to -\zeta$ reverses the signs. So, we just have to consider the case of horizontal and slanting lines through $(0, k\widehat{\tau})$. The sign of the horizontal line, according to Equation 1.3, is $(2k)_{2\omega} > 0$. When k is odd, the sign of the slanting line, according to Equation 1.4, is $(k)_{2\omega} > 0$. When k is even, the sign of the slanting line is $(k + \omega)_{2\omega} < 0$. $\square$

Computer Tie-In: Open the *planar* and *hexagrid* windows on the main program. On *plaid control* panel turn off all the features except *grid* and *hexagrid*. The buttons on the *hexagrid* window control the lines of various masses and capacities. By clicking on these lines you can see them displayed in the *planar* window.

1.5 LIGHT POINTS

First Definition: We say that an *intersection point* on a grid line is the intersection of that grid line with a slanting line. Suppose that γ is a grid line and $v \in \gamma$ is the intersection of γ with a slanting line σ. We call v *light* (with respect to these two lines) if and only if

$$\text{sign}(\sigma) = \text{sign}(\gamma), \qquad \text{mass}(\sigma) < \text{capacity}(\gamma). \tag{1.5}$$

Otherwise we call v *dark*. These features of an intersection point will be called its *shade*.

Second Definition, Horizontal Case: Reflection in any horizontal grid line preserves the grid and maps the slanting lines of negative slope to the slanting lines of positive slope. Therefore, every intersection point on a horizontal line is contained in 2 slanting lines. The two slanting lines containing a horizontal intersection point have the same slope up to absolute value. Let $\epsilon_0, \epsilon_+, \epsilon_- \in \{-1, 1\}$ respectively denote the sign of the horizontal line through v, the positive slanting line through v, and the negative slanting line through v. We call v *light* if and only if

$$\epsilon_0 = \epsilon_- = \epsilon_+. \tag{1.6}$$

Note that this definition just uses the signs.

Second Definition, Vertical Case: Thanks to the fact that $P + Q = 2$, every vertical intersection point v lies on a positive slanting line and a negative slanting line. The sum of the absolute values of the slopes of these slanting lines is 2. Making the same definitions as in the horizontal case, we say that v is light if and only if

$$\epsilon_\infty = \epsilon_- = -\epsilon_+. \tag{1.7}$$

Here ϵ_∞ is the sign of the vertical line through v. Again, this definition just uses the signs.

We need to take special care for double points. We will just describe the situation for the second definition, because (as we prove below) the first definition is equivalent.

Midline Case: Suppose that v is the midpoint, within a block, of a horizontal grid line. In this case v lies on 4 slanting lines. The sign of the slanting line of slope $\pm P$ coincides with the sign of the slanting line of slope $\mp Q$, and so v is reckoned light or dark whether we used the P-lines or the Q-lines in the definition. We count v either as 2 light points or as 2 dark points.

Corner Case: When v lies on the vertical edge of a block, and we consider v as a point on a vertical edge, we have $\epsilon_\infty = 0$ and so Equation 1.7 never holds. Thus, v is reckoned as dark. If v is considered as a point on a horizontal segment, then we use Equation 1.6 to determine whether v is light or dark. In this case we have $\epsilon_+ = \epsilon_-$ automatically.

Lemma 1.2 *Both definitions assign the same shade to an intersection point.*

Proof: To prove this result, we define the signed mass of a positive slanting line just as in the negative case. Let $v = (x, y)$ be a horizontal intersection point. Suppose that v is an intersection point that lies on slanting lines of slope $\pm P$. The case when v lies on slanting lines of slope $\pm Q$ is similar. The signed capacity of the horizontal line through v is $\kappa = (4py)_{2\omega}$. By symmetry it suffices to consider the case when $\kappa > 0$. The signed capacities of the slanting lines through v are given by

$$\mu_\pm = (2py \mp 4p^2 x)_{2\omega}. \tag{1.8}$$

Note that $\mu_+ + \mu_- \equiv \kappa$ mod 2ω.

If v is light according to the first definition, then $0 < \mu_- < \kappa < \omega$. But then μ_+ and $\kappa - \mu_-$ both lie in $(-\omega, \omega)$ and are congruent mod 2ω. This forces $\mu_+ + \mu_- = \kappa$. But then all the signs agree. Hence v is light according to the second definition. Conversely, if v is light according to the second definition, then $\mu_\pm \in (0, \omega)$. Again, this forces $\mu_+ + \mu_- = \kappa$. But then $0 < \mu_- < \kappa$ and v is light according to the first definition.

Now consider the vertical case. We will suppose that $v = (x, y)$ is an intersection point contained on a slanting line of slope $-P$ and a slanting line of slope Q. The other case is similar. We define κ as above. Again, we will consider the case when $\kappa > 0$. The other case has a similar treatment. The quantity μ_- computes the signed mass of v relative to the P-line and we define

$$\lambda_+ = (2py - 4pqx)_{2\omega}. \tag{1.9}$$

The sign of the Q line is the sign of λ_+. We have that $\mu_- - \lambda_+ \equiv \kappa$ mod 2ω.

If v is light according to the first definition, then $0 < \mu_- < \kappa < \omega$. Since $\lambda_+ \in (-\omega, \omega)$ and $\lambda_+ \equiv \mu_- - \kappa$ we must have $\lambda_+ < 0$. Hence v is light according to the second definition. Conversely, if v is light according to the second definition, then we have $\mu_-, \kappa \in (0, \omega)$ and $\lambda_+ \in (-\omega, 0)$. The congruence condition above forces $\kappa = \mu_- - \lambda_+$. But then $\mu_- < \kappa$ and v is light according to the first definition. $\square$

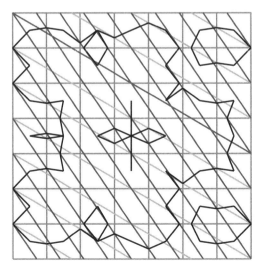

Figure 1.3: The light points in $[0, 7]^2$ with respect to $p/q = 2/5$.

Figure 1.3 shows the light intersection points inside the square $[0,7]^2$ for the parameter $2/5$. For each unit square S in this region, we have connected the center of S to the light points on ∂S. Notice that some interesting curves seem to emerge. Notice also that there seems to be a small amount of junk, in the form of little loops, hanging off these curves. The junk occurs wherever there are two light points in a single edge. The one exception appears to be the case when the edge intersects the vertical midline (right in the center of the picture) but in this case the one light point on each relevant horizontal edge is counted twice.

Computer Tie-In: Open up the *planar window* on the main program. On the *plaid model* control panel, turn off all the features except *grid, light points*, and *polygons*. You can see how the light points determine the polygons. If you turn on the *limit capacity* feature below the *plaid model* control panel, you can just plot the grid lines having capacity less than a number you can control with the arrow keys.

1.6 TRANSVERSE DIRECTIONS FOR THE LIGHT POINTS

Given a slanting line L, let y be the Y-intercept of L. Consider the quantity

$$\delta(L) = ((q-p)y)_{2\omega}. \qquad (1.10)$$

We leave L undirected if $\delta(L) = 0$. We direct L downward if $\delta(L) < 0$ and upward if $\delta(L) > 0$. Let Λ denote a grid line and suppose that $p = L \cap \Lambda$ is a light point. We attach an arrow to p which points perpendicular to Λ and makes a positive dot product with the oriented version of L. We call such a point a *transverse directed light point*.

Since each light point lies on two slanting lines, it remains to show that the transverse directions are well defined. Before proving this, we work out an example.

Example: Again, the parameter is $2/5$. Figure 1.4 shows an example of 3 light points and the slanting lines that contain them. These light points are contained in the boundaries of two shaded unit integer squares. The light point in the corner, at $(0,2)$, is attached to the horizontal line $y = 2$. A comparison with Figure 1.3 shows that there are no other light points in the boundaries of these two shaded squares.

The relevant slanting lines have Y-intercepts $1, 2, 4, 5$. We have $q - p = 3$. This leads to values

$$(3 \times 1)_{14} = +3, \quad (3 \times 2)_{14} = +6, \quad (3 \times 4)_{14} = -2, \quad (3 \times 5)_{14} = +1.$$

In this example, our rule makes the slanting lines with these Y-intercepts point up,up,down,up respectively. These directions consistently define transverse directions for the 3 light points.

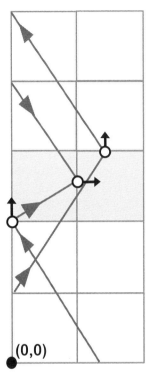

(0,0)

Figure 1.4: Determining the transverse directions.

Lemma 1.3 *Each light point has a well-defined transverse direction.*

Proof: Let L be a slanting line through $(0, y)$. Let $\sigma(L)$ be the sign of L and let $\delta(L)$ be as in Equation 1.10. Let $U = 2py$ and $V = (q - p)y$.

$$\sigma(L) = -\text{sign}(U_{2\omega}), \qquad \delta(L) = +\text{sign}(V_{2\omega}), \qquad U + V = \omega y.$$

From this equation, we see that $\sigma(L) = \delta(L)$ iff y is odd. Put another way, if the Y-intercepts of slanting lines L_1 and L_2 have the same parity, then $\delta(L_1) = \delta(L_2)$ if and only if L_1 and L_2 have the same sign.

Let $\zeta = (x, y)$ be a light point and let L_1 and L_2 be the two slanting lines through ζ. Let y_1 and y_2 be the two Y-intercepts. When ζ is horizontal, L_1 and L_2 have the same type and opposite slopes. Hence $y_1 + y_2 = 2y \in 2\mathbf{Z}$. In the vertical case L_1 has slope $\pm P$ and L_2 has slope $\mp Q$. Since $P + Q = 2$, we have $y_2 - y_1 = \pm 2x \in 2\mathbf{Z}$. In both cases, y_1 and y_2 have the same parity.

In the horizontal case L_1 and L_2 have the same sign. Hence, both point up or both point down. In the vertical case L_1 and L_2 have opposite signs. Hence, both point left or both point right. $\square$

1.7 MAIN DEFINITION

A *unit integer square* is a unit square whose vertices have integer coordinates. Such squares are bounded by grid lines. We call the edges of such a square *unit segments*. We say that a light point on the boundary of a unit integer square Q is *relevant* if it is the only light point on the edge of Q that contains it. We have the following theorem.

Theorem 1.4 *Let Q be any unit integer square. Relative to any even rational parameter, Q has either 0 or 2 relevant light points in its boundary. In the latter case, one of the transverse directions points into Q and the other points out.*

We will assume Theorem 1.4 for now. We will deduce this result as an immediate consequence of Theorem 8.2. The proof of Theorem 8.2 does not depend on any of the structural results worked out in Part 1, so our argument is not circular. Here is the definition that Theorem 1.4 allows us to make.

Main Definition: The *oriented plaid model* takes as input the even rational parameter p/q and assigns either the empty set or a directed edge e_Q to each unit integer square Q. The empty set is assigned when there are 0 relevant light points in the boundary of Q. Otherwise, the directed edge e_Q connects the centers of the sides of Q which contain the relevant light points and crosses these sides in the same direction as the transverse directions. The *unoriented plaid model* is the assignment of unoriented edges obtained by forgetting the orientations. When we don't care whether the model has orientations or not, we will just say the *plaid model*.

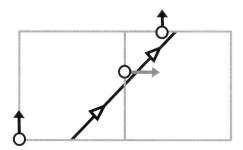

Figure 1.5: Determining the directed edges from the light points.

Figure 1.5 shows the directed edges assigned to the two shaded squares from Figure 1.4. Figure 1.6 shows the unoriented plaid model edges in $[0, 7]^2$. Figure 1.5 fits into the left side of the third row from the bottom in Figure 1.6.

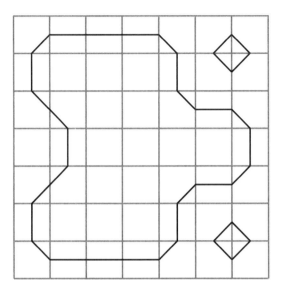

Figure 1.6: The assigned edges inside $[0,7]^2$ for $p/q = 2/5$.

We define a *plaid polygon* to be a connected component of the union of edges in the plaid model. From the definition of the plaid model, these really are embedded polygons. Moreover, abutting edges have consistent orientations. Hence, each plaid polygon is consistently oriented. Figure 1.6 shows the plaid polygons in the block $[0,7]^2$ for the parameter $2/5$. We have not drawn in the orientations, but based on Figure 1.5 we can say that the big polygon is oriented clockwise. It turns out that the two small polygons are oriented counterclockwise. We let PL_A denote the union of all the oriented plaid polygons associated to the parameter $A = p/q$.

Computer Tie-In: Open the *planar* window as above. On the *plaid model* control panel, turn on *grid, (-)slanting lines, directions* and *oriented tiles*. You can see how the tile directions are determined by the directions given to the slanting lines. (The program doesn't draw the directions for the *(+) slanting lines*.) If you want to see the unoriented plaid polygons, turn *polygons* on and *oriented tiles* off. Whether you are displaying the oriented or unoriented polygons, use the big red arrow keys on the upper right portion of the control panel (just below the *reset* button) to cycle through the different blocks of the model. If you change parameters in the midst of all this, you should press the *reset* button to reset your location in the plane.

Chapter Two

Properties of the Model

2.1 CHAPTER OVERVIEW

In this chapter we derive some basic properties of the plaid model. As usual, we fix an even rational parameter $A = p/q \in (0,1)$. We use the basic quantities defined in §1.2. For instance, $\omega = p + q$.

In §2.2 we work out the symmetries of the plaid model. First we deal with the unoriented model and then we consider the oriented model. In §2.3 we prove the technical lemma that each unit integer segment contains exactly 2 intersection points. The work in this section reveals the nice geometric way that the slanting lines intersect each unit integer square.

In §2.4, the highlight of the chapter, we establish the following result: Within each block, there are exactly 2 lines of capacity k for each even $k \in [0, \omega]$. Moreover, within the block, each line of capacity k has exactly k light points on it (when double points are appropriately counted). This is the reason why we use the word *capacity* to describe the numbers assigned to the grid lines. We remind the reader that the material in §2.4 is not needed for Parts 2-4 of the monograph.

In §2.5 we establish a subtle additional symmetry of the plaid model.

2.2 SYMMETRIES

We first deal with the unoriented plaid model.

The Symmetry Lattice: We fix some even rational parameter p/q. Let $\omega = p + q$ as above. Let $L \subset \mathbf{Z}^2$ denote the lattice generated by the two vectors

$$(\omega^2, 0), \qquad (0, \omega). \tag{2.1}$$

We call L the *symmetry lattice*.

As we have already mentioned, the grid lines of mass 0 divide the plane into $\omega \times \omega$ blocks. We define $[0, \omega]^2$ to be the *0th block*. The lattice L acts on the plane in such a way that it preserves the division into blocks. We define the *fundamental blocks* to be $B_0, ..., B_{\omega-1}$, where B_0 is the 0th block and

$$B_k = B_0 + (k\omega, 0). \tag{2.2}$$

The union of the fundamental blocks is a fundamental domain for the action of L. We call this union the *fundamental domain*.

Translation Symmetry: The assignment of signed masses and signed capacities to the integer points on the y-axis is invariant under translation by $(0, \omega)$. Likewise the assignment of signed capacities to the integer points along the x-axis is invariant under translation by $(\omega, 0)$. Finally, if we translate by the vector $(0, \omega^2)$, each slanting line is mapped to another slanting line of the same type whose Y-intercept has been translated by either $P\omega^2$ or $Q\omega^2$, both of which are multiples of ω. Hence, the unoriented plaid model is invariant under the symmetry lattice L. The picture in any block is translation equivalent to the picture in a fundamental block.

Rotational Symmetry: Reflection in the origin – i.e., rotation by π radians about the origin – preserves all the masses and capacities and reverses all the signs, and hence is a symmetry of the unoriented plaid model. Combining this with the translation symmetry, we see that reflection in the center of the fundamental domain is also a symmetry of the unoriented plaid model. This center is the center of the block $B_{\omega-1}/2$. Likewise, the reflections in the midpoints of the sides of the fundamental domain are symmetries of the unoriented plaid model.

Reflection Symmetry: Reflection in the Y-axis preserves the signs of the horizontal grid lines, reverses the signs of the vertical grid lines, preserves the signs of the slanting lines, and reverses the slopes of the slanting lines. From the second definition of the light points, we see that reflection in the Y-axis permutes the light points and hence is a symmetry of the unoriented plaid model. Combining this fact with the rotational symmetry just discussed, we see that reflection in the X-axis is also a symmetry of the unoriented plaid model. (Of course, this can also be worked out directly.)

Combining the reflection and translation symmetry, we see that reflection in the horizontal midline of a block is a symmetry of the unoriented plaid model. This explains the bilateral symmetry one sees in the pictures of the plaid polygons. For use in the proof of Theorem 2.3 below, we mention a fine point about this horizontal reflection: It preserves the type P horizontal light points, preserves the type Q horizontal light points, and interchanges the type P vertical light points with the type Q vertical light points. What is going on is that the horizontal light points are on two slanting lines of the same type and the vertical light points are on two slanting lines of differing types. (The type is determined from the slanting lines of negative slope.)

Special Dihedral Symmetry: The special case we now consider is not so important for us, but it does give us a chance to test the theory. Combining the reflection symmetry and the rotation symmetry of the block $B_{(\omega+1)/2}$, we see that $B_{(\omega+1)/2}$ always has 4-fold dihedral symmetry. Figure 2.1 shows B_{10} with respect to the parameter $5/14$.

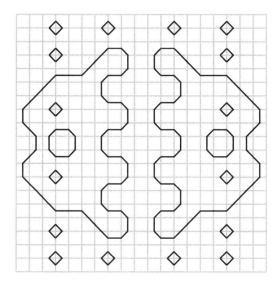

Figure 2.1: B_{10} for the parameter $5/14$.

Now we turn to the oriented plaid model. We want to see that every symmetry discussed above either globally respects the directions of the edges or globally reverses them.

Lemma 2.1 *We have the following symmetry:*

1. *Translation by $(\omega, 0)$ reverses the directions.*

2. *Translation by $(\omega^2, 0)$ respects the directions.*

3. *Rotation in the origin respects the directions.*

4. *Reflections in the coordinate axes respect the directions.*

5. *Reflection in the horizontal midline of a block reverses the directions.*

Proof: For all statements, let ρ be the given transformation of the plane. Let L_0 be a slanting line. Let $L_1 = \rho(L_0)$. Let y_j be the Y-intercept of L_j for $j = 0, 1$. We set $\sigma(\rho) = 1$ if ρ carries an upwards parametrization of L_0 to an upwards parametrization of L_1. Otherwise we set $\sigma(\rho) = -1$. In each case below we will show that $\delta(L_1) = \pm\delta(L_0)$, where the sign does not depend on the choice of L_0. If the sign matches $\sigma(\rho)$ then ρ globally respects the directions of the edges in the oriented plaid model. In the other case, ρ globally reverses the directions of the edges in the oriented plaid model.

1. For Statement 1, we have $y_1 = y_0 + \omega$. Hence
$$(q - p)y_1 - (q - p)y_0 \equiv \omega \mod 2\omega.$$
But then $\delta(L_1) = -\delta(L_1)$. Also $\sigma(\rho) = 1$.

2. For Statement 2, we have

$$y_1 - y_0 = \omega^2 P = 2p\omega \equiv 0 \mod 2\omega.$$

This time we have $\delta(L_1) = \delta(L_2)$ and $\sigma(\rho) = 1$.

3. For Statement 3, we have $y_1 = -y_0$ and hence $\delta(L_0) = -\delta(L_1)$. This time $\sigma(\rho) = -1$.

4. For Statement 4, we first consider the reflection in the Y-axis. We have $\delta(L_0) = \delta(L_1)$ and $\sigma(\rho) = 1$. Hence ρ globally respects the directions. The case of reflection in the X-axis follows from Statement 3 and the case of reflection in the Y-axis.

5. Statement 5 is a consequence of Statements 1 and 4.

This completes the proof. $\square$

When we consider the oriented plaid model modulo translation symmetry, we have 2ω fundamental blocks rather than ω fundamental blocks. We will discuss this below in more detail.

2.3 THE NUMBER OF INTERSECTION POINTS

In this section we will consider the number of intersection points on a unit integer segment. We first discuss double points. When a double point appears at the endpoint of an edge, we count it once. When a double point occurs in the interior of the edge – and this only happens at the center of a horizontal edge – we count it twice. This agrees with our conventions in §1.5, and it also makes sense from the geometric perspective taken in the proof of the next lemma.

Lemma 2.2 *Each unit integer segment contains 2 intersection points.*

Proof: We give an arithmetic argument in the vertical case and a geometric argument in the horizontal case.

Vertical Case: Let e be a vertical unit integer segment. Let m be the x-intercept of the line containing e. The type P points on e have the form $n + \alpha$ where $n \in \mathbf{Z}$ and $\alpha \in [0,1)$ is the fractional part of Pm. The type Q points on e have the form $n + \beta$ where $n \in \mathbf{Z}$ and $\beta \in [0,1)$ is the fractional part of Qm. Note that $p - q$ and ω are relatively prime. Hence

$$Pm - Qm = \frac{2(p-q)m}{\omega} \in \mathbf{Z}$$

iff ω divides m. Likewise $Pm \in \mathbf{Z}$ iff ω divides m. So, either $\alpha = \beta = 0$ or $\alpha \neq \beta \in (0,1)$. In the first case, we have double points at both endpoints of e and so we count 2 intersection points in e. In the second case, we get exactly 2 points in the interior of e.

Horizontal Case: Our argument here is indirect. Let S be a unit integer square and let Γ denote the intersection of S with the union of all slanting lines. There are two basic facts.

- Each horizontal intersection point in ∂S is the endpoints of 2 segments of Γ having slope set $\{+P, -P\}$ or $\{+P, -P\}$.

- Each vertical intersection point in ∂S is the endpoints of 2 segments of Γ having slope set $\{+P, -Q\}$ or $\{+Q, -P\}$.

From these facts, and from the fact that $P + Q = 2$, we see that Γ is a union of quadrilaterals (often immersed and possibly degenerate) that are each inscribed in Q with one vertex on each side. This remains true even in the degenerate case, with our conventions.

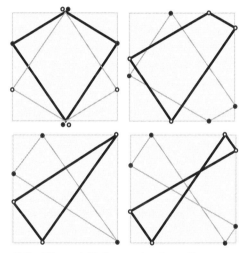

Figure 2.2: The set Γ drawn to show its components: 4 examples.

Figure 2.2 shows the 4 main possibilities for the geometry of these quads. These examples are all taken from the case when the parameter is $p/q = 2/5$. The number of quads of Γ is the same as the number of intersection points on any edge of ∂S. Hence, the edges of ∂S have the same number of intersection points, provided that these intersection points are counted with our conventions. We know already that the vertical edges have 2 intersection points. Hence, the horizontal edges do as well. $\square$

Remark: The interested reader can probably tell from Figure 2.2 that once the values P and Q are fixed, these inscribed quads fit inside a 1-parameter family. In the rational case we are considering, we only see finitely many representatives of the family when we look at the plaid model.

2.4 THE MEANING OF CAPACITY

Now we come to a more subtle result which suggests the hierarchical nature of the plaid model and explains our use of the word *capacity*.

Theorem 2.3 *Let B be any block in the plaid model relative to the parameter p/q. Let $\omega = p + q$. Then*

1. *For each even $k \in [0, \omega]$ there are 2 lines in $\mathcal{H}$ and 2 lines in $\mathcal{V}$ which have capacity k and intersect B.*

2. *Each such line contains exactly k light points that lie in B.*

Computer Tie-In: Open the *planar* window on the main program. On the *plaid model* control panel, turn off all the features except *grid* and *light points*. Turn on the *limit capacity* option below the *plaid model* control panel and use the arrow keys to set the capacity limit to something small, like 2 or 4. Notice the number of light points on each line. Now use the arrow keys underneath the *reset* button to see how the light points move as you change blocks. (The *reset* button nicely scales the picture in the *planar* window. You should use it frequently.) I recommend using the parameter 21/34 and setting the capacity to 4.

We will prove Theorem 2.3 through a series of smaller lemmas. By symmetry, it suffices to take B to be a fundamental block and to consider only lines of positive sign.

Lemma 2.4 *Statement 1 of Theorem 2.3 is true.*

Proof: Given the periodicity of the capacity labels, it suffices to prove this when B is the 0th block. We will prove the result for $\mathcal{H}$. The result for $\mathcal{V}$ has virtually the same proof. We are simply trying to show that there are exactly 2 integer values of y in $[0, \omega]$ such that $4py = \pm k$ mod 2ω. Writing $k = 2h$, we see that this equation is equivalent to $2py = \pm h$ mod ω. This has 2 solutions mod ω because $2p$ is relatively prime to ω. $\square$

Lemma 2.5 *Theorem 2.3 holds in the vertical case.*

Proof: Let L be the nontrivial intersection of a vertical line of signed capacity $+k$ with B. We will show that there are $k/2$ light points of type P on L. The same result holds with Q in place of P, by the reflection symmetry discussed in §2.2. Since $k > 0$ we are in the case of Lemma 2.2 where $\alpha \neq \beta \in (0, 1)$. In other words, there are no double points. So, the $k/2$ light points of type P are distinct from the $k/2$ light points of type Q, and we get a total of k as desired.

Let $S \subset \mathbf{Z}$ denote the set of Y-intercepts of slanting lines of slope $-P$ which intersect L. The set S is a run of ω consecutive integers and hence contains all the congruence classes mod ω, and each one exactly once.

The number of light points of signed mass μ on L equals the number of $y \in S$ such that

$$(2py + \omega)_{2\omega} = \mu. \tag{2.3}$$

Since ω is odd and $2p$ and 2ω are both even, the left-hand side of Equation 2.3 is odd. Since $2p$ is relatively prime to ω, two different values of y give the same value in Equation 2.3 if and only if they are congruent mod ω. Since there are ω possible values for the left hand side of Equation 2.3 and ω possible inputs mod ω, we see by the pigeonhole principle that Equation 2.3 has a unique solution for $y \in S$ when $\mu \in (0, k)$ is odd. This shows that there is exactly one light point of signed mass μ on L for each $\mu = 1, 3, ..., (k-1)$. This gives us $k/2$ light points of type P on L. $\square$

Before we launch into the horizontal case, we give two preliminary results which will streamline the proof.

Congruence Principle: Given any $I = \{i_1, ..., i_k\} \subset \mathbf{Z}$ we define the translated set $I + n = \{i_1 + n, ..., i_k + n\}$. Let $N > 0$ be some positive integer. (In the application we will have $N = 2\omega$.) We say that I is N-*full* if I contains every congruence class mod N.

Here is the principle. Suppose that I_1 and I_2 are subsets of integers such that $I_1 \cup I_2$ is N-full. Let $I_1' = I + N_1$ and $I_2' = I_2 + N_2$ where $N_1 + N_2 \equiv 0$ mod N. Then $I_1' \cup I_2'$ is also N-full.

To see this, let $I_j'' = I_j' + 1$ for $j = 1, 2$. Then $(I_1'' \cup I_2'') = (I_1' \cup I_2') + 1$. From this we see that $I_1' \cup I_2'$ is N-full if and only if $I_1'' \cup I_2''$ is N-full. Applying this fact repeatedly, we reduce to the case where $N_1 = kN$ for some k and $N_2 = 0$. In this case, the result is obvious.

Geometric Principle: We also mention a geometric fact which will streamline our proof. Suppose λ_1 and λ_2 are two slanting lines of slope P and $-Q$ respectively. Let y_j be the Y-intercept of λ_j. If $y_2 - y_1 = j\omega$ for some even j then $\lambda_1 \cap \lambda_2$ lies on the vertical boundary of a block. If j is odd then $\lambda_1 \cap \lambda_2$ lies on the vertical midline of the block. This is to say that this intersection point lies at the midpoint of the intersection of some horizontal line with the block. These results use the fact that the difference in the slopes, namely $P - (-Q) = P + Q$, is 2, and that the blocks have width ω.

Now we are ready for the horizontal case.

Lemma 2.6 *Theorem 2.3 holds in the horizontal case.*

Proof: We proceed somewhat as in the vertical case. Let B be the mth fundamental block. Here $m \geq 0$. Let L be the nontrivial intersection of a horizontal line of signed capacity $+k$ with B. It suffices to prove that L has exactly 2 light points of mass μ for each $\mu \in \{1, 3, ..., k-1\}$. Let $S(s)$ denote the set of Y-intercepts of slanting lines having slope s which intersect L. Let $S = S(P) \cup S(-Q)$. The light points on L of mass μ correspond to integers $y \in S$ satisfying Equation 2.3.

Let us work out the structure of the set S. Let y_0 be the Y-intercept of L. The set $S(P)$ is a run of $2p + 1$ consecutive integers going down from $y_0 - 2mp$ and $S(-Q)$ is a run of $2q + 1$ consecutive integers going up from $y_0 + 2mq$. In the special case when $m = 0$ we can see that S is a run of $2\omega + 1$ consecutive integers. Hence S is 2ω-full. Moreover, the only two numbers in S with the same congruence mod 2ω are $\min S(P)$ and $\max S(-Q)$. Since $2pm + 2qm = 2\omega m$, the set S is 2ω-full by the congruence principle. Moreover, if two distinct numbers in S are congruent mod 2ω, then these points are either $\{\min S(P), \max S(-Q)\}$ or $\{\max S(P), \min S(-Q)\}$. Again by the geometric principle, the slanting lines through such points intersect L at one endpoint or the other.

Since S is 2ω-full and $\mu \in (0, k)$ is odd, Equation 2.3 has at least two solutions $y, y' \in S$ such that $y' = y + j\omega$ for some odd j. If the corresponding light points are not distinct, then they must have different types and moreover must lie at the intersection point of the two relevant slanting lines. But then the geometric principle says that they correspond to the double point at the center of L. By convention, we count these as two light points.

Suppose that there are more than 2 numbers in S which solve Equation 2.3. Since $2p$ is relatively prime to ω, we see that amongst any 3 solutions to Equation 2.3 we must have 2 that are congruent to each other mod 2ω. But then the corresponding light points are both the same endpoint of L and we count the two of them just once, by convention. Once we pair off two solutions like this, giving us one light point, we still have one or two more solutions, which give us the second light point. So, in all cases, we get exactly 2 light points of mass μ. □

2.5 A SUBTLE SYMMETRY

The result here helps later on to explain the order 8 dihedral symmetry of our vertical pixelated spacetime diagrams. This is a minor technical result which plays little role in the monograph. I recommend that the reader skip this section on the first pass.

We call two light points on the same vertical grid line a δ-*pair* if they lie δ units apart from each other. We call a vertical grid line δ-*paired* if all the light points on it come in δ-pairs. Here δ must be the same for all pairs.

Lemma 2.7 *For each even $\kappa \in \{2, ..., \omega - 1\}$ there exists some $\delta < 1/2$ and a δ-paired grid line of capacity κ.*

Proof: Let $y_1, y_3, ..., y_{\kappa-1} \in (0, \omega)$ be the Y-intercepts whose associated signed mass sequence is $1, 3, ..., \kappa - 1$. Let $L(P, j, m)$ be the slanting line of slope $-P$ whose Y-intercept is $y_j + \omega m$. Likewise define $L(P, j, m)$.

Consider the vertical line $V(m)$ that is

$$f(m) = \frac{m\omega}{Q - P} = m \times \frac{(p + q)^2}{2q - 2p}$$

units from the Y-axis. $V(m)$ contains all the intersection points

$$L(P, j, 0) \cap L(Q, j, m), \qquad j = 1, 3, ..., \kappa - 1.$$

Let $I = \{0, ..., 2q - 2p\}$. Since p/q is an even rational, there are $2q - 2p$ numbers in the set

$$J = \bigcup_{m \in I} f(m) \subset [0, \omega^2]$$

and these numbers are equidistributed mod ω. Hence every point in $[0, \omega]$ is within $\omega/(4q - 4q)$ of some point of J mod ω.

Interpreting the construction above geometrically, we see that there is some choice of $m \in (0, 2q - 2p)$, some fundamental block B, and vertical grid line of capacity κ such that $V(m)$ is $\delta < \omega/(4q - 4p)$ from V. But then the light points on V come in pairs, and the distance between partner points is

$$(Q - P)\delta < (Q - P) \times \frac{\omega}{4q - 4p} = 1/2.$$

Hence V is δ-paired for some $\delta < 1/2$. $\square$

We define the *pair center* of a δ-pair to be the center of mass of the two light points. We call a δ-paired line *endpoint centered* (respectively *midpoint centered*) if all the pair centers are at endpoints (respectively midpoints) of unit integer segments.

Lemma 2.8 *Let $\delta < 1$. A δ-paired vertical grid line is either of endpoint centered or of midpoint centered.*

Proof: Let V be the δ-paired grid line. Since the distances between light points of the same type on V are integers, in every δ-pair on V the same type light point is on top. Hence, the distance between pair centers is always an integer. Reflection in the X-axis permutes the pairs and hence their centers. Hence, the pair centers are symmetrically located about the X-axis. This means that the distance from the pair centers to the X-axis is either always an integer or always a half-integer. $\square$

Chapter Three

Using the Model

3.1 CHAPTER OVERVIEW

In this chapter we explore some consequences of the results in the previous chapter, especially Theorem 2.3. The main theme in the chapter is that Theorem 2.3 gives a way to extract information from the geometry of the low capacity lines.

In §3.2 we prove that, relative to the parameter p/q, the 0th block always contains a polygon whose projection onto the X-axis has diameter at least $(p+q)/2$.

In §3.3 we elaborate on the theme in §3.2 to show how to extract increasingly fine scale information about the plaid polygons. In §3.4 we explain how to augment the idea in §3.3 to make it more useful. In §3.5 we show how the ideas from §3.3 sometimes explain why the plaid model looks similar at different rational parameters.

We will take up some of the themes in this chapter again in Part 5 of the monograph, when we have assembled more information about the plaid model. We remind the reader that nothing in this chapter is needed for Parts 2-4 of the monograph.

3.2 THE BIG POLYGON

Let p/q be an even rational parameter. We continue using the notation from §1.2. In particular, $\omega = p + q$ and $Q = 2q/\omega$.

Theorem 3.1 *Let $B = [0, \omega]^2$ be the 0th block with respect to p/q. There exists a unique plaid polygon $\Gamma \subset B$ whose projection onto the x-axis has diameter at least $\omega/2$. Moreover, Γ has bilateral symmetry with respect to reflection in the horizontal midline of B.*

Proof: Let L be the segment obtained by intersecting B with the horizontal line through the point $(0, \widehat{\tau})$. According to Lemma 1.1, the line extending L has signed capacity $+2$ and any slanting line through $(0, \widehat{\tau})$ has signed mass $+1$. Hence the point $z_1 = (0, \widehat{\tau})$ is a light point on L.

Consider the slanting line Λ of slope $-Q$ through the point $(0, \widehat{\tau} + \omega)$. Λ also has signed mass $+1$. Since $Q > 1$, the line Λ intersects L at a point

$z_2 = (x, \widehat{\tau})$. Here

$$x = \frac{\omega}{Q} = \frac{\omega^2}{2q} = \frac{\omega}{2} \times \left(1 + \frac{p}{q}\right) > \frac{\omega}{2} + \frac{1}{2}.$$

Now we know that L has two light points z_1 and z_2 spaced more than $\omega/2 + 1/2$ units apart. For $j = 1, 2$ let I_j be the unit integer segment on L which contains z_j and let m_j be the midpoint of I_j. Since L has capacity 2, Theorem 2.3 tells us that z_1 and z_2 are the only light points on L. This means that a plaid polygon can only intersect L at the points m_1 and m_2. Plaid polygons are continuous loops transverse to the grid lines, so any plaid polygon intersects the line containing L an even number of times. At the same time, no plaid polygon intersects ∂B. Hence, a plaid polygon intersects L an even number of times. So, the plaid polygon which contains m_1 also contains m_2. Let Γ be this polygon.

The distance from m_1 to m_2 is greater than $\omega/2 - 1/2$. Hence, the projection of Γ to the X-axis is greater than $\omega/2 - 1/2$. Since this projection of Γ is always an integer or a half-integer, we see that the projection of Γ has diameter at least $\omega/2$.

The vertical grid lines of capacity 2 which intersect B are evenly spaced with respect to the vertical midline of B. One of these lines, V, lies less than $\omega/2$ from the Y-axis. But then Γ must intersect V. Reasoning as above, we see that Γ intersects V twice and no other plaid polygon in B interects V. This shows that Γ is the unique plaid polygon in B whose projection to the X-axis has diameter at least $\omega/2$. In particular, $\Gamma = \Gamma'$, where Γ' is obtained by reflecting Γ in the horizontal midline of B. Hence Γ has bilateral symmetry. $\square$

We call the polygon Γ from Theorem 3.1 *the big polygon*. Figure 3.1 shows 2 examples.

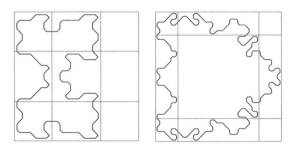

Figure 3.1: The picture for 5/18 and 14/31.

Computer Tie-In: Open up the main program. When the main program is running, open up the *planar window* and the *plaid model* control panel. Turn off all the features except *first polygon* and *grid*. Turn on the *limit capacity* option and set it to 2. Repeatedly click on the *random* button to generate lots of examples of the first polygon and how it interacts with the capacity 2 lines.

3.3 HIERARCHICAL INFORMATION

In this section we elaborate on the themes of the last two sections, showing how to get increasingly fine scale information using more of the model. We will work with the oriented plaid model because it gives more information. We fix an even rational parameter which is implicit in everything.

We say that a *directed point* on the boundary of a rectangle is a point on this boundary equipped with an arrow which either points out of the rectangle or into it. These are the kind of directions we assigned to the light points in §1.6. We say that a *decorated rectangle* is a rectangle with finitely many directed points on its boundary.

A *connector* between two directed points is a continuous arc which joins two directed points in such a way that one of the directions points out of the rectangle and the other one points into it. We insist that the arc stays in the interior of the rectangle except at its endpoints.

Given a decorated rectangle, we say that a *connection pattern* associated to the rectangle is a finite union of connectors which uses all the points, such that no two connectors cross. A necessary and sufficient condition for a connection pattern to exist is that there are the same number of inward pointing points as outward pointing points. The necessity of this condition is obvious. The sufficiency can be proved by induction. The inductive step involves pairing off an inward point to an adjacent outward point. Note that more than one connection pattern might exist relative to the same decorated rectangle. The first two rectangles in Figure 3.2 admit a unique connection pattern up to isotopy. The last two (which have the same decoration) do not.

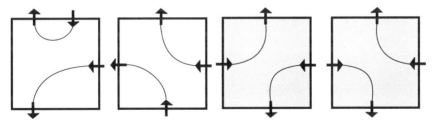

Figure 3.2: Connection patterns.

We call a decorated rectangle *unambiguous* if there is exactly 1 connection pattern associated to it, up to isotopy. Otherwise we call the rectangle *ambiguous*. The Rectangle Lemma from the previous section tells us that at least half the rectangles in the partition associated to Γ_R are unambiguous.

Suppose now that we have some even rational parameter p/q and some block B. We say that an *integer rectangle partition* of B is a partition of B into rectangles whose sides are contained in the grid lines. The rectangles of the partition need not meet edge to edge. Say that a *special point* is the midpoint of a unit integer segment that has exactly one light point. We direct the special point so that the direction points the same way as the transverse direction associated to the light point. The decoration of R is simply the union of all the directed special points contained in the boundary of R.

Since every plaid polygon that enters R must also exit R we see that the decoration of R admits a connection diagram. We call the integer rectangle partition *unambiguous* if every decorated rectangle is unambiguous. When we have an unambiguous integer rectangle partition, we can piece together the individual connection diagrams to produce a finite union of piecewise smooth loops. We call the resulting union of polygons the *coarse loops* associated to the partition.

Before stating the next result, we remind the reader that the *Hausdorff distance* between two sets S_1 and S_2 is the infimal ϵ such that each S_i is contained in the ϵ-tubular neighborhood of S_j. This distance turns the set of compact subsets of $\mathbf{R}^2$ into a metric space. We define the *mesh* of an integer rectangle partition to be the maximum diameter of a rectangle in the partition.

Theorem 3.2 *Let $\mathcal{R}$ be some unambiguous rectangle partition of B, which has mesh ϵ. Then there is a bijection between the set of coarse loops associated to $\mathcal{R}$ and the plaid polygons in B which are not confined to individual rectangles in $\mathcal{R}$. Loops which correspond under this bijection are at most ϵ apart in the Hausdorff metric.*

Proof: Let R be any rectangle of $\mathcal{R}$. Let S_1 denote the connection pattern for R used to create the coarse loops. Let S_2 be the union $\gamma \cap R$, taken over all plaid polygons γ which intersect ∂R. Then S_2 is also a connection pattern for R. Since R is unambiguous, S_1 and S_2 are isotopic relative to an isotopy which does not move the endpoints. In particular, there is a canonical bijection between the connectors of S_1 and the connectors of S_2. These individual bijections piece together to give a bijection between the plaid polygons in B which are not confined to single rectangles of $\mathcal{R}$ and the coarse loops. Within each rectangle, corresponding connectors are within ϵ of each other. Hence, the Hausdorff distance between corresponding loops is at most ϵ. $\square$

An example says a thousand words. For any even K we can get an integer rectangle partition of any block B by simply taking the rectangles cut out by the grid lines of capacity at most K. We take $p/q = 5/12$ and consider the 0th block. We take the partition corresponding to $K = 6$ and then we add one small segment of capacity 8 to improve the picture. For this partition, we can usually determine the connection pattern from the undirected light points. We have shaded the rectangles in which we need to use the directions to figure it out.

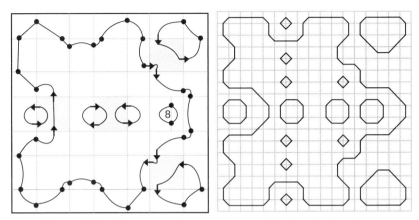

Figure 3.3: The coarse loops compared to the plaid polygons.

From Figure 3.3 we can see that B contains 7 plaid polygons which intersect our partition. One of these polygons is the big polygon from Theorem 0.7. The remaining polygons require the higher capacity lines for us to see them. The right side of Figure 3.3 shows the actual plaid polygons in B for this parameter. The smallest plaid polygons are confined to single rectangles of the partition.

Computer Tie-In: I have implemented the above construction at least for rectangle grids made from all the lines less than a given capacity. Go into the *Wild* directory and run the program there. Kill all the windows except the *main control panel* and the *plaid model* window. Pick the parameter like $5/12$. Set the *plot method* to *basic* and the □ *grid depth* to 3. Press *go*. This will show a figure like Figure 3.3 but without the capacity 8 segment added in. Warning: The program in this directory is not well documented and is not easy to use.

3.4 A SUBDIVISION ALGORITHM

One difficulty with Theorem 3.2 is that it might not be so easy to come up with an unambiguous rectangle partition. One ambiguous rectangle spoils the whole enterprise. Here we explain how to potentially improve an ambiguous rectangle partition using a recursive subdivision algorithm.

Subdivision Algorithm: Suppose that $\mathcal{R}$ is an ambiguous rectangle partition. We choose the first ambiguous rectangle R we see and divide R by the lowest capacity line which intersects R in its interior. We alternate between horizontal and vertical cuts to keep the rectangles fairly square. We could take the initial partition $\mathcal{R}$ to be the set of rectangles cut out by the lines of capacity at most K. By varying K the algorithm (if successful) would produce a sequence of unambiguous partitions at varying scales.

Example: We will illustrate this with an example. We start with the parameter 38/161 (chosen more or less at random) and some block B (also chosen more or less at random) and the same partition using grid lines of capacity at most 6.

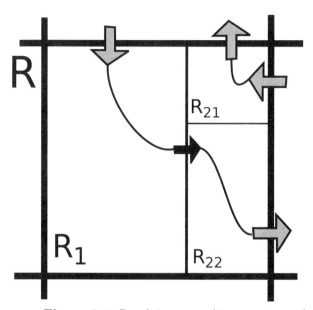

Figure 3.4: Resolving an ambiguous rectangle.

The big rectangle R in Figure 3.4 belongs to the partition. Notice that it is ambiguous. However, we can subdivide it into 3 smaller rectangles which are unambiguous. The smaller rectangles all involve lines of capacity greater than 6. We first add the lowest capacity vertical line which intersects the interior of R. This line intersects R in a segment having one light point, and divides R into one unambiguous rectangle R_1 and one ambiguous rectangle R_2. We then take the lowest capacity horizontal line that intersects the interior of R_2. This line intersects R_2 in a segment which has no light point and divides R_2 into the two unambiguous rectangles R_{11} and R_{12}.

Typically the subdivision algorithm quickly terminates, resulting in an unambiguous partition. However, if we carefully choose the blocks and the parameters, we can make it go as long as we like. The idea is to pick a block B which contains large polygons Γ_1 and Γ_2 that come very close to each other and have the same orientation. It often happens that there is some other block B' and a large polygon Γ' that looks very much like a translation of $\Gamma_1 \cup \Gamma_2$. For such blocks, we need to look at a very fine scale in order to decide whether we are looking at one polygon or two.

This phenomenon will seem more natural later when we consider the 3-dimensional incarnation of the plaid model. The different blocks are horizontal slices of a cube filled with polyhedral surfaces. The plaid polygons are the slices of the surfaces. If we pick two slices close to, but on different sides of, a saddle-type critical point for the height function, we can arrange the situation we are talking about.

Figure 3.5 shows an example where it takes much longer to resolve an ambiguous rectangle. We have not completed the resolution here. The two shaded rectangles are still ambiguous. We have somewhat indicated the order in which the subdivision takes place by the thickness of the lines. The thicker lines are added first.

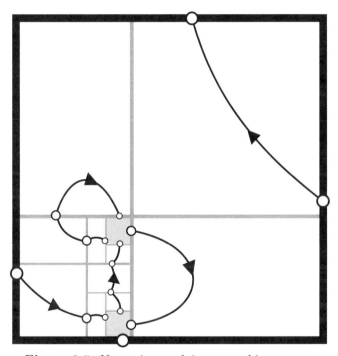

Figure 3.5: Not quite resolving an ambiguous rectangle.

Since the plaid model is consistent, the subdivision algorithm always terminates in a completely unambiguous pattern of rectangles. However, if we end up with the final grid of 1 by 1 squares we haven't really obtained useful coarse information about the plaid model.

Computer Tie-In: To see the recursive subdivision algorithm in action, follow the same steps as in the previous computer tie-in, but this time set the *plot method* to *recursive*. Incidentally, if you set the *plot method* to *perfect* you can see a plot of the plaid polygons. You can experiment around and see this algorithm work for other parameters and other initial rectangle partitions.

3.5 COMPARING DIFFERENT PARAMETERS

Theorem 3.2 sometimes explains why the plaid model looks about the same on a large scale when computed relative to two different parameters. Figure 3.6 shows the plaid polygons in the 0th block for the two different parameters $p_1/q_2 = 5/12$ and $p_2/q_2 = 12/29$. Notice that there is a very good correspondence between the 7 largest polygons on each side.

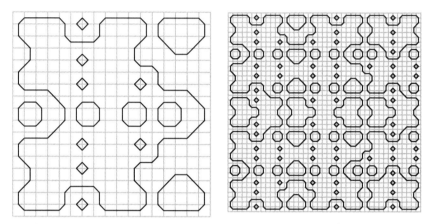

Figure 3.6: The plaid polygons in the 0th block for 5/12 and 12/29.

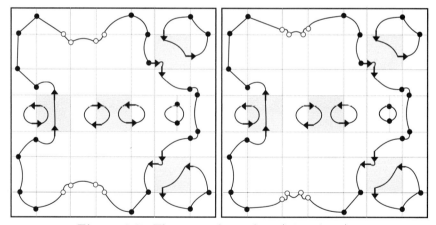

Figure 3.7: The coarse loops for 5/12 and 12/29.

Here is an explanation. The left side of Figure 3.3, the picture of the coarse loops associated to a particular integer rectangle partition, looks pretty similar when we compute it for 12/29 in place of 5/12. Figure 3.7 shows the two pictures side by side. The white points highlight the minor topological difference between the two pictures.

The two parameters are related to each other in two nice ways:

- They are quite close together. Indeed, they are consecutive continued fraction approximations of $\sqrt{2} - 1$.

- The two ratios $\tau_1/\omega_1 = 5/17$ and $\tau_2/\omega_2 = 17/41$ are also quite close together.

Thanks to Lemma 1.1, what is important for the *relative location* of the low capacity grid lines in a block is the ratio τ/ω. For instance, if $\tau/\omega \approx 1/7$ then the two lines of capacity 2 are about $1/7$ of the way from the top and bottom of each block and the two lines of capacity 4 are about $2/7$ from the top and bottom of the block. The same goes for the low mass slanting lines and hence for the light points on the low capacity grid lines.

For our two parameters, the edges of the partition are almost exactly in the same positions, and so are the light points on them. This explains the similarities between the two pictures. The highlighted points in the figure are very close to the grid point, so small changes can cause them to cross over to the other side. That explains the differences between the two figures.

Our idea gains in strength when we consider an infinite sequence of parameters, such as

$$\frac{5}{12} = [0:2:2:2], \qquad \frac{12}{29} = [0:2:2:2:2], \qquad \frac{29}{70} = [0:2:2:2:2:2], \quad \cdots.$$

Here we are listing the continued fraction expansion so as to explain the pattern. The picture relative to the odd terms in the sequence looks like the left side of Figure 3.7 and the picture relative to the even terms looks like the right side of Figure 3.7. The highlighted points converge to the relevant grid point as the parameter tends to $\sqrt{2} - 1$.

When we pick some very complicated parameter, say $[0 : 2 : ... : 2]$ with 999 consecutive (2)s, the big plaid polygons have more than 2^{1000} sides, but still the picture on the left-hand side of Figure 3.7 predicts roughly what the biggest plaid polygons look like in the 0th block.

So far, we have just been talking about what happens in the 0th blocks. If we want to line up other blocks, we would have to be more careful. We will explain how to line up the other blocks in §5.6.

Chapter Four

Particles and Spacetime Diagrams

4.1 CHAPTER OVERVIEW

In this chapter we will begin the process of making a 3D interpretation of the plaid model. The idea is to group together certain of the light points and think of them as instances of 1-dimensional worldlines rather than as a succession of points. We fix an even rational parameter p/q and we use the notation from §1.2. In particular, $\omega = p + q$.

In §4.2 we will explain a different notion of adjacency for the $\omega \times \omega$ blocks dividing up the plaid model. It is not useful for us to say that blocks are adjacent when they share an edge. In our notion, which we call *remote adjacency*, two remotely adjacent blocks are separated by $\omega \widehat{\tau}$. Here $\widehat{\tau} \in (0, \omega)$ is such that $2p\widehat{\tau} \equiv 1 \bmod \omega$. This notion of remote adjacency will tell us which particles to group together.

In §4.3 we say what it means for two horizontal light points in remotely adjacent blocks to be different instances of the same particle. Our point of view is that, as we cycle through remotely adjacent blocks in order, we are seeing the same particle at different times. We will formalize this idea in the notion of a spacetime diagram. In §4.4 we do the same for the vertical particles.

In §4.5 we show a few pictures of spacetime diagrams and discuss their symmetries.

In §4.6 we will prove a technical lemma, the Bad Tile Lemma, which is very similar in spirit to Theorem 1.4. This result guarantees that the polygons produced in the next chapter are all embedded.

I strongly suggest that the reader skip the proofs in §4.6 on the first pass.

Notational Convention: In our construction, we find it easier to identify sets of the form S and $S + (m\omega^2, 0)$ in the plaid model. Here m is an integer. The reason we do this is that translation by $(\omega^2, 0)$ is a symmetry of the oriented plaid model. With our convention we don't have to keep translating sets back into our fundamental domain discussed in §2.2. Any point of set we discuss has a representative modulo symmetry in the fundamental domain and this is good enough for us.

4.2 REMOTE ADJACENCY

Define

$$B_k^\pm = [k\omega, k\omega + \omega] \times [0, \pm\omega]. \tag{4.1}$$

Here $B_0^+, ..., B_{\omega-1}^+$ are the fundamental blocks. In the unoriented plaid model, the two blocks B_k^+ and B_k^- have exactly the same picture in them. In the oriented model, these blocks have the same picture except that all the orientations have been reversed. We write

$$B_k^\pm \to B_{k-\widehat\tau}^\mp. \tag{4.2}$$

We call the two blocks involved in this relation *remotely adjacent*.

Since $\widehat\tau$ is relatively prime to ω, the cycle

$$B_0^+ \to B_{-\widehat\tau}^- \to B_{-2\widehat\tau}^+ \to B_{-3\widehat\tau}^- \to \cdots \tag{4.3}$$

encounters every one of the blocks defined above. Here we are using the convention of identifying any block with its representative in the fundamental domain. We call the cycle in Equation 4.3 the *fundamental cycle*. It has length 2ω.

Relative Motion: Now we are going to describe a more radical kind of translation that does not respect the oriented plaid model. If we have some point z in some block B, we let $[z]$ denote the translate of z in the 0th block B_0^+. That is, $[z] = T(z)$ where T is the translation that carries B to B_0^+. We make this convention so that if we have a sequence of points $z \to z' \to z'' \to ...$ we can watch the relative motions of the points $[z] \to [z'] \to [z''] \to ...$ all inside B_0^+. We emphasize that the point $[z]$ is not (necessarily) a light point in the plaid model when z is. The point $[z]$ is just kind of a marker to help us track the relative position of z.

Computer Tie-In: To see the relative motion in action, open the *planar* window on the main program. On the *plaid model* control panel, turn off all the features except *grid* and *light points*. Turn on the *limit capacity* option below the *plaid model* control panel and use the arrow keys to set the capacity limit to (say) 4. Pick a nice parameter like 17/38. Push the *rescale* button, then use the red arrow keys to move through the fundamental cycle of blocks. Watch the relative motion of the light points.

4.3 HORIZONTAL PARTICLES

Now we explain the relative motion of the horizontal light points as we move from block to block through the fundamental cycle. It will turn out that we can group together certain of these light points into a kind of moving particle.

Lemma 4.1 *Suppose $B \to B'$. Let H be a horizontal grid line which intersects B. Let H' be the horizontal line that intersects B' in the same relative position. Let $z \in H \cap B$ be an intersection point. Suppose z has type P (respectively Q) and is not on the right (respectively left) edge of B. Then there is an intersection point $z' \in B' \cap H$ of the same type and shade as z such that $[z'] - [z] = (P^{-1}, 0)$ (respectively $([z'] - [z]) = (-Q^{-1}, 0)$). If z is a light point, then the transverse direction of z' is also the same as the transverse direction of z.*

Proof: We will treat the P case. The Q case has the same proof. Let L be the slanting line of slope $-P$ that contains z. There is a slanting line L^* of slope $-P$ through the right edge of B and the difference in Y-intercepts between L and L^* is at least 1. Hence z is at least P^{-1} from the right edge of B. Define

$$z' = z + (-\widehat{\tau} + P^{-1}, \pm \omega). \qquad (4.4)$$

The sign in the second coordinate is chosen according to the parity of B in such a way that $z' \in B' \cap H'$. By construction, z' still lies on a slanting line L' of slope $-P$. The difference between the Y-intercepts of L and L' is $-2P\widehat{\tau} + 1 + \omega$. This number is even and congruent to $0 \bmod \omega$. Hence it is also $0 \bmod 2\omega$. Hence L and L' have the same signed mass. Hence z is a light point iff z' is. In case z is a light point, the transverse directions of z and z' are determined by the directions of L and L'. These lines have the same sign and their Y-intercepts differ by an even number. From the argument in Lemma 1.3 these lines have the same direction. Hence, z and z' have the same transverse direction when they are light points.

There are vectors U and U' such that

$$[z] = z + U \in B_0^+, \qquad [z'] = z' + U' \in B_0^+, \qquad U' - U = (\widehat{\tau}, \mp \omega). \ (4.5)$$

We compute that

$$[z'] - [z] = \big(U + (\widehat{\tau}, \mp \omega) + z'\big) - \big(U + z\big) =$$

$$(\widehat{\tau}, \mp \omega) + (-\widehat{\tau} + P^{-1}, \pm \omega) = (P^{-1}, 0).$$

This completes the proof. □

Horizontal Particles: If z and z' are related as in the lemma above, we write $z \to z'$. We call the collection of points in the cycle

$$z \to z' \to z'' \to \cdots \to z^{(2\omega - 1)} \qquad (4.6)$$

a *horizontal particle*. We call each of the points in the cycle an *instance* of the particle. If we go through the fundamental cycle twice and watch the relative positions of the instances of a particle, the points appear to move rightward with speed P^{-1} and leftward with speed Q^{-1}. The reader can see this in action using my computer program.

Horizontal Spacetime Diagrams: The horizontal segment $H \subset B_0^+$ of length ω contains all the points $[z], [z'], [z''], \dots$ These points are not themselves light points, but they are markers indicating the relative motion of the particle within the 0th block. We want to keep track of these markers in a more visually appealing way.

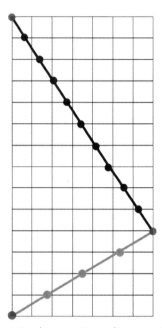

Figure 4.1: A spacetime diagram for a horizontal particle.

Figure 4.1 shows a *spacetime diagram* which records the relative positions of successive points of the particle. The bottom row of the diagram shows the location of $[z]$ on H. The second row shows the location of $[z']$ on H. And so on. All the integer rows in the spacetime diagram are identified with the segment H.

In the example shown in Figure 4.1, the parameter is $2/5$ and the horizontal line has capacity 2. We are tracking the particle whose 0th instance is the light point $(0, 2)$ of mass 1. The top and bottom point are meant to be identified, so that there are $14 = 4 + 10 = 2 \times 7$ instances in total. The slopes of the lines are $P = 4/7$ and $-Q = -10/7$. All other diagrams for the horizontal light points for the same parameter look the same up to translation. We call the slanting lines in the figure *worldlines*.

4.4 VERTICAL PARTICLES

In the vertical case, the situation is superficially different because each vertical line intersects at most one fundamental block. So, we consider a family of vertical segments $\{V_k\}$ such that V_k intersects the kth block in the fundamental cycle in the same relative position. Each vertical segment is the intersection of a vertical grid line with the relevant block. We call these segments *vertical block segments*. We write $V_0 \to V_1 \to V_2, \dots$ to indicate this.

Lemma 4.2 *Let $B \to B'$ be fundamental blocks. Let $V \to V'$ where V is a vertical block segment of B and V' is a vertical block segment of B'. Let $z \in V$ be an intersection point of type P (respectively Q) which is more than 1 unit from the top (respectively bottom) of B. Then there is an intersection point $z' \in V'$ of the same type and shade so that $[z'] = [z] + (0, 1)$ (respectively $[z'] = [z] - (0, 1)$). If z is a light point then the transverse directions of z and z' are the same.*

Proof: We will consider the type P case. We let

$$z' = z + (0, 1) - (\hat{\tau}\omega, \pm\omega). \tag{4.7}$$

Let L be the slanting line of slope $-P$ through z and let L' be the slanting line of slope $-P$ through z'. The difference in the Y-intercepts of these two lines is

$$-P\hat{\tau}\omega + 1 \pm \omega = -2p\hat{\tau} + 1 \pm \omega.$$

This number is even and congruent to 0 mod 2ω. Hence it is congruent to 0 mod 2ω. The rest of the proof is as in the horizontal case. For the type Q case, everything is the same except that we have $-(0, 1)$ in Equation 4.7. What makes this work is that $-2q\hat{\tau} - 1 \pm \omega = -2\omega\hat{\tau} + 2p\hat{\tau} - 1 \pm \omega$ is congruent to 0 mod 2ω. $\square$

Involving Reflection: Lemma 4.2 is not the final word because we have not dealt with the situation when our light points are within 1 unit of the top or bottom of a fundamental block. We consider the type P case in detail. We omit the completely symmetric details for the type Q case.

Let ρ denote reflection in the horizontal midline of the block B'. Let's assume that z is a light point to save words. We define z' by the following steps:

1. Let z_1 be the point from Equation 4.7. This point now lies less than one unit above the top of B'. It is a light point of type P and has the same transverse direction as z. The formula for the relative position of z_1 is not the same however.

2. Let $z_2 = z_1 - (0, \omega)$. By symmetry z_2 is a light point of type P and transverse direction opposite z and z_1.

3. Let $z' = \rho(z_2)$. By construction z' is a light point having type Q and the same transverse direction as z.

The relative position of z' is such that $[z]$ and $[z']$ are both on the same vertical block segment of the 0th block, and the sum of their distances from the top of the block is 1. Essentially, $[z']$ is the result of bouncing $[z]$ off the top of the block, changing the type, and keeping the transverse direction.

Vertical Particles: We define the *vertical particle* $z \to z' \to z'' \to \dots$. There are ω instances of the particle having type P and ω instances having type Q. When the particle has type P it moves upward in a relative sense and when it has type Q it moves downward in a relative sense. The transverse directions are the same for all instances of the particle. The type changing may seem unnatural at first, but it is exactly the thing needed to make the vertical particles oscillate up and down just like the horizontal particles oscillate left to right.

Vertical Spacetime Diagrams: We can make a spacetime diagram as in the horizontal case. This time we turn the vertical grid lines horizontal and stack them on top of each other. (The left edge in the picture corresponds to the bottom of the blocks.) Figure 4.2 shows the diagram for one of the light points on a vertical line of capacity 2 for the parameter 2/5. As in the horizontal figure, there are $14 = 7 + 7$ instances of the particle in this example. The worldlines in this case always have slope ± 1. As in the horizontal case we identify the top and bottom of the spacetime diagram so that really the domain is a cylinder. What we have is a billiard path of slope ± 1 in this cylinder!

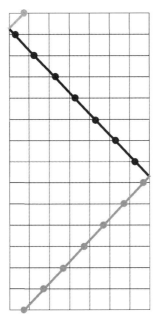

Figure 4.2: A spacetime diagram for a vertical particle.

4.5 SPACETIME DIAGRAMS AND THEIR SYMMETRIES

In the spacetime diagrams above, we just showed one particle at a time. However, below we show all the particles associated to a given grid line at the same time. According to Theorem 2.3, if the grid line has capacity k, there will be k translated copies of the polygonal paths shown in Figures 4.1 and 4.2. Figure 4.3 shows the diagrams for the grid lines of capacity 2. We have used the parameter 2/5 again.

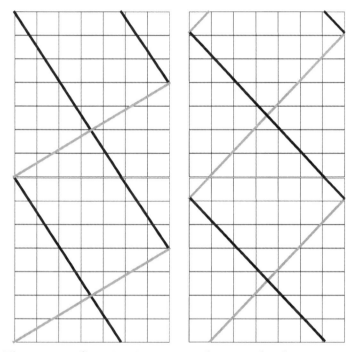

Figure 4.3: Capacity 2 spacetime diagrams for the parameter 2/5.

Computer Tie-In: Open the *planar* and *spacetime diagrams* windows. Click the middle mouse button (or use key-x) on the grid edges in the *planar* window and you will see the spacetime diagram for the particles on the corresponding slice. If you just want to see low capacity diagrams, turn on the *limit capacity* feature beneath the *plaid window* control panel. If you want to systematically explore these diagrams, bring forth the *spacetime diagram* control panel.

The translation $\tau(x, y) = (x, y + \omega)$ is a symmetry of the figure. This symmetry comes from the fact that the oriented plaid model has an orientation-reversing translation symmetry. Figure 4.4 shows the quotient of Figure 4.3 by τ. We call this the *quotient spacetime pictures*.

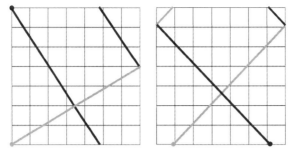

Figure 4.4: The quotient picture.

Figure 4.5 shows two of the four capacity 4 slices for the parameter $2/5$. Again, we are showing the quotient picture. In the horizontal case on the left we have highlighted the places where there are triple intersections. The horizontal spacetime diagrams always have 2-fold rotational symmetry. This rotational symmetry derives from the fact that reflection in the Y-axis is a symmetry of the unoriented plaid model. See §2.2.

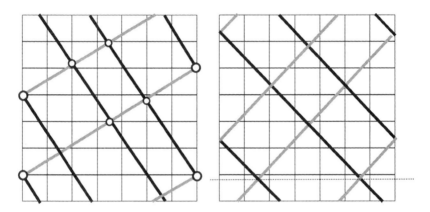

Figure 4.5: Capacity 4 spacetime diagrams for the parameter $2/5$.

The vertical spacetime diagrams always have 2-fold bilateral symmetry across the vertical midline. This symmetry derives from the fact that reflection in the X-axis is a symmetry of the unoriented plaid model. See §2.2.

There is more symmetry in the vertical spacetime diagrams, which we now explain. This extra symmetry only comes up again in the remark at the end of §5.3.

Say that a *special rectangle* is a rectangle whose sides have slope ±1. We have drawn an auxiliary dotted line in Figure 4.5. If we cut the cylinder open along either dotted line, the union of worldlines becomes a union of special rectangles inscribed in the resulting square and having order 8 dihedral symmetry. Moreover, the places where the rectangles touch the dotted line lie on the midlines of unit integer squares in the diagram. That is, their X-coordinates are half-integers. We show that this holds in general.

Recall from §2.5 that a δ-paired vertical line is one in which all the light points on it come in pairs separated by δ. Now that we know about particles we can give a quick corollary to Lemma 2.7.

Lemma 4.3 *For each even $\kappa \in (-\omega, \omega)$ there is some $\delta < 1$ and a midpoint centered δ-paired vertical grid line of capacity κ.*

Proof: The vertical line V from Lemma 2.7 has all the right properties except that it might be endpoint centered. Suppose this happens. We will suppose that all the type P light points are above their partner type Q light points in each pair. The other case has a similar treatment.

Let $a = (\omega + 1)/2$. Let V' be the vertical grid line such that $V \to \ldots \to V'$, where there are a arrows. That is $V' = V - (a\omega\hat{\tau}, 0)$. All the light points of type P have moved a units down and all the light points of type Q have moved a units up and the whole picture is ω-periodic. But then we see that V' is $(1 - \delta)$-paired and midpoint centered. □

Let H be the integer horizontal segment that corresponds to V in our spacetime diagram. The worldlines in the diagram intersect H in pairs of points which are separated by $\delta < 1$. The centers of mass of these pairs lie on the midlines of unit integer squares. The dotted line H^* is either $\delta/2$ units above H or $\delta/2$ units below, depending on whether the relevant worldlines cross above H or below.

When we cut open along H^* the result we get is a union of inscribed special rectangles. We know already that this union has vertical bilateral symmetry. But it is not hard to see that a finite union of special rectangles inscribed in a square has order 8 dihedral symmetry provided that it has bilateral symmetry. This completes the proof.

Computer Tie-In: Repeat the instructions given for the computer tie-in discussed in §4.2 and watch for the δ-paired situation to arise (in a relative sense) on your favorite vertical line as you cycle through the blocks.

4.6 THE BAD TILE LEMMA

In the next chapter we are going to convert our spacetime diagrams into unions of embedded polygons. In this section we prove a technical result which guarantees that the polygons really are embedded.

Say that a *tile* in a spacetime diagram is a unit integer square. Say that a tile in a spacetime diagram Σ is *bad* if there exist exactly two worldlines W_1 and W_2 which intersect T, and each one of these worldlines intersects T in a pair of opposite edges. This situation cannot happen in the vertical case because the worldlines have slope ± 1. We just have to worry about the horizontal case.

Lemma 4.4 (Bad Tile) *There are no bad tiles in a spacetime diagram.*

Remark: I strongly suggest that the reader skip the proof of the Bad Tile Lemma on the first pass through the monograph.

We first discuss some more structure of the spacetime diagram and then we give the proof of the Bad Tile Lemma. Let Σ be a horizontal spacetime diagram. The points where the worldlines intersect the left edges of Σ are all integer points. The reason for this structure is that each left endpoint ζ^* corresponds to a light point ζ which lies in the vertical edge of a block. We call these points *left endpoints*. (The right endpoints have the same property, but we prefer to work with the left endpoints.) Given left endpoints ζ_1^* and ζ_2^* let ζ_1 and ζ_2 be the corresponding light points. Let Δ be the difference in the signed masses of these points, taken in either order.

Lemma 4.5 *Let m be the distance from ζ_1^* to ζ_2^*. Then $\Delta \equiv \pm 2pm$ mod 2ω.*

Proof: We will prove this result in case m is odd. The even case is very similar. For each left endpoint $\zeta_j^* = (0, m_j)$ the corresponding light point ζ has the equation

$$\zeta_j = (-m_j\widehat{\tau}\omega, *), \tag{4.8}$$

where $(*)$ is some integer we do not care about. The y coordinates of ζ_1 and ζ_2 are equal when m is even and differ by $\pm\omega$ if m is odd.

Since ζ_j^* is the intersection of two worldlines, ζ_j lies on all 4 kinds of slanting lines. Let L_j denote the slanting line of slope $-P$ through ζ_j. Let y_j denote the Y-intercept of L_j. The signed mass of ζ_j is given by

$$(2P\widehat{\tau}y_j)_{2\omega}. \tag{4.9}$$

Taking the points in a suitable order, we have

$$\zeta_1 - \zeta_2 = (m\widehat{\tau}\omega, \pm\omega). \tag{4.10}$$

Since L_1 and L_2 have slope $-P = -2p/\omega$,

$$y_1 - y_2 = Pm\widehat{\tau}\omega \pm \omega = 2pm\widehat{\tau} \pm \omega. \tag{4.11}$$

(In the even case this would be the same but without $\pm\omega$.) Hence

$$\Delta = (2p \times (2pm\widehat{\tau} \pm \omega)). \tag{4.12}$$

By definition, $2p\widehat{\tau} \equiv 1$ mod ω. Hence

$$2pm\widehat{\tau} \pm \omega \equiv m \text{ mod } \omega. \tag{4.13}$$

Hence one of the following equations is correct:

$$2pm\widehat{\tau} \pm \omega \equiv m \text{ mod } 2\omega \qquad 2pm\widehat{\tau} \pm \omega \equiv m + \omega \text{ mod } 2\omega$$

Since m and ω are odd, both sides of the first congruence have the same parity whereas the two sides of the second congruence have opposite parity. Hence, the first of the two equations is correct. When we plug this congruence into Equation 4.12 we get the conclusion of the lemma. The sign ambiguity in the conclusion of the lemma comes from the fact that we might have switched the order of the points. □

Proof of the Bad Tile Lemma: Without loss of generality we can consider the case when Σ has positive signed capacity. Consider the intersection $z = W_1 \cap W_2 \in T$. Let ζ_1^* and ζ_2^* be the left endpoints contained in W_1 and W_2 respectively. Let m be the distance between ζ_1^* and ζ_2^*. Let ζ_1 and ζ_2 be the corresponding light points.

When m is even, z lies on a vertical edge of T. In this case, the two worldlines both intersect this edge, and we do not have a bad tile. When m is odd, z lies on the vertical midline of T, as shown in Figure 4.6. This is the case we need to deal with.

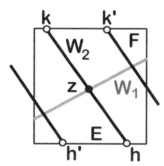

Figure 4.6: Inside a bad tile.

Let E and F respectively be the bottom and top edges of T. If z lies less than $2/P$ from E or F, then both W_1 and W_2 intersect E or F and we do not have a bad tile. So, we can assume that z lies more than $P/2$ from either E or F.

W_2, which has slope $-Q$, intersects E at a point h which is at least

$$1/2 + (1/Q)(P/2) = 1/Q$$

from the left endpoint of E. To avoid a lot of verbiage, we identify E with the unit integer segment in the planar plaid model that contains the corresponding light point, and we identify this light point with h. Let L denote the slanting line of slope $-Q$ through h (in the planar model). Let L' denote the slanting line of slope $-Q$ whose Y-intercept is one less than that of L. Let h' be the intersection of L' with the horizontal grid line containing h. The distance between h and h' is $1/Q$ and h' lies to the left. Hence $h' \in E$ as well. We can make the same construction with respect to the top edge F. This gives us points k and k'.

Let μ_j be the signed mass of ζ_j. We claim that

$$\mu_1 \in (0, \mu_2 + 2p), \qquad \mu_2 \in (0, 2p). \qquad (4.14)$$

We deal with μ_2 first. μ_2 is the same as the signed mass of h. Suppose that h has signed mass $\mu_2 > 2p$. The slanting line of slope $-Q$ through h'

is obtained by taking the slanting line of slope $-Q$ and translating it down 1 unit. Using the formula for signed mass given in §1.4, we see that h' has signed mass $\mu_2 - 2p$. If this quantity is positive, then it lies in $(0, \mu_2)$ and hence h' would be another light point. But then there would be another worldline through h' and this worldline intersects T. This is a contradiction. Hence $\mu_2 \in (0, 2p)$.

Now we deal with μ_1. If $\mu_2 + 2p \geq \omega$ then Σ has capacity less than $\mu_2 + 2p$ because all slices have capacity less than ω. Since ζ_1 is a light point on a grid line of positive sign, we have $\mu_1 \in (0, \mu_2 + 2p)$. Consider the case when $\mu_2 + 2p < \omega$. If $\mu_1 \geq \mu_2 + 2p$ then Σ has capacity at least $\mu_2 + 2p + 1$. (We add 1 because capacities are even and masses are odd.) The same argument as above shows that the point k' has signed mass $\mu_2 + 2p$. This uses the fact that $\mu_2 + 2p < \omega$. But then k' is another light point. We get the same contradiction as before.

By Lemma 4.5, we see that $\pm 2pm$ is congruent to $\mu_1 - \mu_2$. Given our bounds on μ_1 and μ_2 in Equation 4.14 we see that $2pm$ is congruent to a number in $(-2p, 2p) \bmod 2\omega$. Hence $(2pm)_{2\omega} \in (-2p, 2p)$. Since $P = 2p/\omega$, we can divide this last equation by 2ω to get

$$(Pm/2)_1 \in (-P/2, P/2). \tag{4.15}$$

Here $(x)_1$ denotes the result of taking x and subtracting the nearest integer.

The distance from z to the left edge of Σ is m. Moreover, z lies on a line of slope P which starts out at the integer point ζ_1^*. But then the second coordinate of z is exactly $(Pm/2)_1$ away from the nearest horizontal edge of T. This says that z comes within $P/2$ of a horizontal edge of T. This is a contradiction. $\square$

Here is a special case of the Bad Tile Lemma that will be useful later on.

Lemma 4.6 *Suppose that Σ is a horizontal spacetime diagram of capacity at most $2p$. If two non-parallel worldlines W_1 and W_2 intersect a tile T in Σ then either $W_1 \cap W_2$ lies in a vertical edge of T or else both W_1 and W_2 intersect one of the horizontal edges of T.*

Proof: In this case, Equation 4.14 is forced without any assumptions on the location of the point z. Using this equation we see that z comes within $P/2$ from the nearest horizontal edge of the tile T. But then W_1 and W_2 both intersect this horizontal edge. $\square$

Chapter Five

Three-Dimensional Interpretation

5.1 CHAPTER OVERVIEW

The purpose of this chapter is to explain the 3-dimensional interpretation of the plaid model. We fix an even rational parameter p/q and we use the notation from §1.2. In particular, $\omega = p + q$. This chapter builds on what we did in the previous chapter.

In §5.2 we will stack the blocks on top of each other in such a way that remotely adjacent blocks appear actually adjacent to each other in the stack. We stack these blocks at integer heights and then fill in the space between the stacks to create a rectangular solid. (Working with the unoriented model we get a cube and working with the directed model we get two cubes stacked on top of each other.) Once this is done, we will reinterpret the spacetime diagrams constructed in the previous chapter as integer XZ and YZ slices of our solid.

In §5.3 we show how to modify the spacetime diagrams constructed in §4.3 and §4.4 so that they are unions of embedded loops, much like the plaid polygons. We call this modification *pixelation*. The Bad Tile Lemma is what guarantees that the polygons are embedded. Once this is done, we have a cubical solid filled with disjoint embedded polygons in all the integer coordinate slices.

In §5.4 we will show that the plaid model construction and the pixelation processes are compatible with each other.

In §5.5 we use the compatibility of all our constructions to create polyhedral surfaces which simultaneously interpolate between the plaid polygons and the pixelated spacetime diagrams. We view these surfaces as spacetime diagrams for the plaid polygons. We call these surfaces *spacetime plaid surfaces*.

In §5.6 we indulge in some discussion and speculation. Nothing after §5.2 is used in Parts 2-4 of the monograph.

5.2 STACKING THE BLOCKS

Let $\omega = p+q$ as usual. The 2ω fundamental blocks all lie in the plane, but we can cut them out of the plane (so to speak) and stack them so that remotely adjacent blocks are consecutive integer slices of the rectangular solid

$$\widehat{\Omega} = [0, \omega]^2 \times [0, 2\omega]. \tag{5.1}$$

To be more precise, the block $B^{\pm}_{-k\tau}$ is identified with the horizontal slice $[0, \omega]^2 \times \{k\}$ for $k = 0, ..., 2\omega - 1$. We are using the extra dimension to make remotely adjacent slices actually adjacent.

To make the picture nicer, we think of $\widehat{\Omega}$ as a Euclidean orbifold. First of all, the face $Z = 0$ is identified by translation to the face $Z = 2\omega$. Second, the remaining faces are mirrored. These identifications guarantee that the bottom slice and the top slice are also spaced 1 apart in the orbifold. They also capture the "bounce" properties of the spacetime diagrams. The horizontal spacetime diagrams described above are the integer XZ slices of $\widehat{\Omega}$. Likewise the vertical spacetime diagrams are the integer YZ slices of $\widehat{\Omega}$.

When we work with the unoriented plaid model, we take $\widehat{\Omega}$ to be the 2-fold quotient of the above domain under the symmetry $(x, y, z) \rightarrow (x, y, z + \omega)$. The result is an $\omega \times \omega \times \omega$ cube which has the same kinds of identifications on its boundary.

5.3 PIXELATED SPACETIME DIAGRAMS

Technically speaking, the worldlines in the spacetime diagrams are just guides for the eye to follow the particle. However, now we enhance our view of the plaid model so that the worldlines become part of it.

Let Σ be a spacetime diagram. By Lemma 2.2 each unit integer segment of Σ intersects at most 2 worldlines. Let T be a tile of Σ. We call an edge of T *relevant* if it intersects exactly one worldline. Now we make the following definition.

1. If T has exactly 2 relevant edges we join the midpoints of these edges by a segment. We call this segment a *connector*.

2. If T has 4 relevant edges, then we pair the edges of T according to which of the two worldlines they intersect. We draw connectors between the midpoints of edges which correspond to the same worldline.

Note that there are always an even number of relevant edges, because each worldline enters and exits T. In Case 2 there must be exactly 2 worldlines because otherwise we would have more than 4 intersection points. Finally, the connectors in Case 2 do not cross, thanks to the Bad Tile Lemma.

The union of all the connectors gives us a family of embedded polygons. We call the process of replacing the spacetime diagram by the connectors *pixelation*.

The hand-drawn Figure 5.1 shows the pixelation of Figure 4.4. (In the figure caption, we let κ stand for the capacity of the slice.) Since we are really working in a cylinder, with the top and bottom identified, we have taken the liberty of vertically translating the picture to make it look as nice as possible. We will always make such translations. We have superimposed (the translated version of) Figure 4.4 so as to make the pixelation process more clear. As in the plaid model construction, we associate the left and right endpoints in the horizontal case to the horizontal edges of the tiles that contain it.

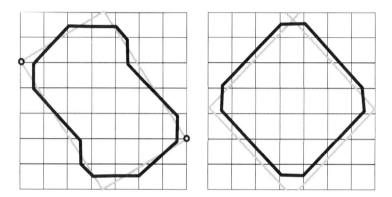

Figure 5.1: Pixelated spacetime diagrams for $\kappa = 2$ and $p/q = 2/5$.

The pixelation rules allow for the particles to merge and lose their individual identities. In Figure 5.1 each pixelated path is a closed and centrally symmetric loop in the orbifold. The pixelation process seems rather delicate, but we will see in the next chapter that there is actually a much more robust description.

Figure 5.2 shows the pixelated versions of Figure 4.5.

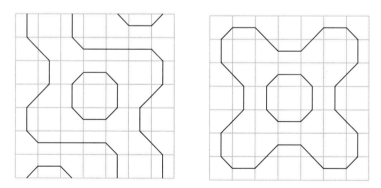

Figure 5.2: Pixelated spacetime diagrams for $\kappa = 4$ and $p/q = 2/5$.

Figure 5.3 shows the capacity 10 diagrams for the parameter $9/16$.

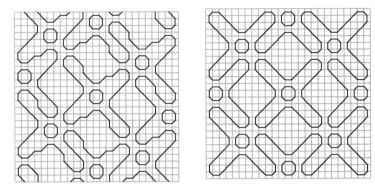

Figure 5.3: Pixelated spacetime diagrams for $\kappa = 10$ and $p/q = 9/16$.

It turns out that relative to the parameter p/q, the horizontal and vertical pixelated spacetime diagrams associated to the capacity κ are always isotopic when $\kappa \leq 2p$. The cutoff value of $2p$ also arises in the proof of the Bad Tile Lemma. See the remark at the end of §7.6.

Remark: All the symmetries we see in these pictures follow from the symmetry discussed at the end of §4.5. In particular, the vertical pixelated spacetime diagrams always have order 8 dihedral symmetry. See §5.6 for more about this.

Computer Tie-In: To see the pixelated spacetime diagrams in action, repeat the instructions for the computer tie-in in §4.5. This time open the *spacetime* control panel and turn on the *pixelated H* and *pixelated V* features. If you want to see these diagrams without the worldlines present, turn off the *worldlines* feature. If you really want to explore the various features, open up the *document* window and click on the *?* box in the *spacetime* control panel.

5.4 TILE COMPATIBILITY

We say that a *tile* in $\widehat{\Omega}$ is a unit integer square contained in one of the unit integer coordinate slices, together with all the connectors in it. A tile in an XY plane is compatible with a time in an XZ plane, in the following sense. If two tiles share an edge, then the midpoint of this edge is incident to a connector in one of the tiles if and only if it is incident to a connector in the other tile. The reason is just that the common edge is a horizontal edge of a unit integer square, and this edge either has a single light point or not. In other words, the pixelation is compatible with the definition of relevant edges used to define the plaid model. The same compatibility holds for tiles in an XY plane and in an YZ plane which are adjacent.

Lemma 5.1 *The tiles in the XZ plane are compatible with the tiles in the YZ plane. That is, if two such tiles share a common edge, then the midpoint of this edge lies in one connector if and only if it lies in the other.*

Proof: We will first describe the 2-dimensional reason behind the compatibility and then we will interpret it in a 3-dimensional way. Suppose that B and B' are blocks, with $B \to B'$. Let H and V respectively be horizontal and vertical grid lines which intersect B. Let H' and V' be corresponding horizontal and vertical grid lines in B'. We write $H \to H'$ and $V \to V'$ to indicate that we think of these as single lines moving through time. Let H_W (respectively H_E) denote the portion of H lying to the west (respectively east) of V.

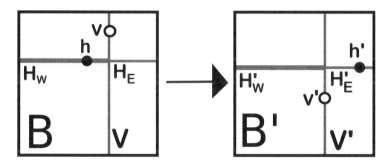

Figure 5.4: A horizontal particle crossing a vertical line.

Suppose that there is one more light point on H_W than there is on H'_W. This means that there is a horizontal particle $h \to h'$ such that $h \in H_W$ and $h' \in H_E$ and no other horizontal particle has crossed over the vertical line $V \to V'$. Figure 5.4 shows the situation.

Consider the 4 rectangular components of $B - (H \cup V)$. Each plaid polygon intersects the boundary of each such component an even number of times. Hence, there are an even number of light points on the boundary of each such component. The same goes for the 4 components of $B' - (H' \cup V')$. For this reason, there must be some vertical particle $v \to v'$ which crosses $H \to H'$ either from north to south or from south to north. Figure 5.4 shows the north to south case.

Going to the 3D picture, the worldline corresponding to $h \to h'$ lies in some XZ slice and crosses the edge E that corresponds to the moving intersection point $(H \cap V) \to (H' \cap V')$. The edge E lies in 2 adjacent XZ-tiles, one of which is the tile T_H shown in Figure 5.5.

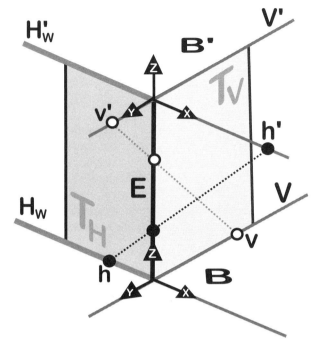

Figure 5.5: Three-dimensional picture.

At the same time, the worldline corresponding to $v \to v'$ lies in some YZ slice and crosses E. This worldline intersects the two adjacent YZ tiles, one of which is the tile T_V shown in Figure 5.5. No other worldlines cross E. The two tiles T_H and T_V are adjacent and both have connectors which go through the midpoint of E. This is the compatibility.

Every time T_H has a connector that involves E we must have the situation described above, and we get the desired connector in T_V. The same goes if we switch the roles of V and H. $\square$

Gear Structure: Say that two plaid polygons *mesh* if they have edges which cross the midpoints of adjacent unit integer segments on the same grid line. In the above proof, we might have also said that the number of points on the boundary of each rectangle having out-pointing transverse directions must equal the number of light points on the boundary having in-pointing transverse directions. This gives another kind of compatibility: The edges where two polygons mesh are pointing in the same direction across the common grid line that they cross. We call a set of disjoint loops with this property a *gear structure* because, if these polygons were turned into flexible gears which interlocked whenever a mesh occurred, they could still all spin at once. We will take up this in more detail in §5.6.

5.5 SPACETIME PLAID SURFACES

Now we know that all the tiles in our construction are compatible. Let's explore the consequences. Consider a unit integer cube in the spacetime picture. On each face of the cube we have a tile. Since these tiles are completely compatible with each other across edges, the union of the connectors on the surface of the cube is a union of closed embedded polygons. Figure 5.6 shows some examples which arise frequently. In these examples, the black edges are parts of plaid polygons. So, we are looking down on the 3D model and the XY slices face us. We have curved a few of the edges to get a better picture.

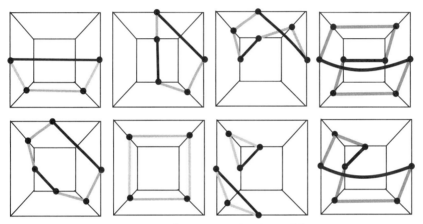

Figure 5.6: Some typical cubical fillings.

Computer Tie-In: On the main program, open the *planar* and *cube* window. Click the middle mouse button over a square in the *planar* window and see the cube whose bottom face is that square. If you drag the mouse over the cube you can change the perspective and see a nice 3D effect. You can use the red arrow keys beneath the *reset button* to go through the fundamental cycle and see how the front face of the cube at any time predicts the plaid connector in the next block.

Each polygon on the surface of a unit integer cube bounds a disk inside the cube. We can fill each disk in by coning to the center of mass of its vertices. The produces a union of polyhedral disks inside the unit cube. We call this union a *filling* of the cube.

If we put in fillings for every unit integer cube, the result is a finite union of pairwise disjoint polyhedral surfaces. The pixelated spacetime diagrams and the plaid polygons are all just slices of what we call *plaid surfaces* in the spacetime interpretation of the plaid model.

Basic Properties: Let Σ be a plaid surface. Σ is orientable because it is a closed surface embedded in $\boldsymbol{R}^3$. By construction Σ is tiled by polygons, each of which is a filling of a unit integer cube. Because the unit cubes fit together 4 around an edge, the polygons in the tiling of Σ meet 4 around a vertex. That is, the 1-skeleton of the tiling is a 4-valent graph.

We can 3-color the edges of the tiling of Σ according to the kind of slice that contains the edge. Around each vertex of the tiling the colors go $ABAB$, where A and B are two of the colors. The set of edges of the same color forms a union of cycles.

- The union of XY edges is the set of plaid polygons contained in Σ.

- The union of XZ edges is the set of polygons in horizontal pixelated spacetime diagrams contained in Σ.

- The union of YZ edges is the set of polygons in vertical pixelated spacetime diagrams contained in Σ.

All together Σ contains a finite number of embedded loops, each having one of 3 colors. These loops intersect each other transversely, and loops of the same color are disjoint. A case-by-case analysis shows that these polygons have at most 8 sides. The right-hand side of Figure 5.6 shows the only 8-sided polygon that can occur.

It is impossible for an XY loop (i.e., a plaid polygon) in Σ to have more than one edge in one of the tiles. The reason is that the two plaid edges in a tile correspond to different blocks of the planar plaid model. See §5.6 for more about the nature of the curves.

Computer Tie-In: My javascript program, discussed at the end of the introduction, draws pictures of the plaid surfaces for smallish parameters. Some of these pictures are quite intricate and beautiful. The program has its own documentation when opened. The program, incidentally, also displays the cube fillings discussed above. In fact, this program does a better job of giving the feel of the 3D plaid model than the main program does.

Examples: The two examples we present here are just a tiny slice of what you can see on the computer program. For each parameter p/q, there is distinguished plaid surface. The distinguished plaid surface contains the big polygon from Theorem 0.7. We call this surface $\Sigma(p/q)$. We first show $\Sigma(1/2)$. It is convenient to work with the unoriented model, which is just the 2-fold quotient of the oriented model. In the unoriented version, there are 3 fundamental blocks, and the plaid polygons are shown in Figure 5.7.

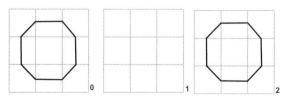

Figure 5.7: The three blocks for the parameter $1/2$.

$\Sigma(1/2)$ is the only surface in this case. It is a topological sphere which is perhaps best described as starting with a Rubik's cube and sanding down the corners. Figure 5.8 shows a topologically accurate planar picture. The thick lines make the plaid polygons. The thin lines come from the pixelated worldlines. If we forget the coloring of the lines, the surface has a symmetry group of order 48.

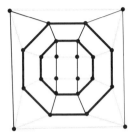

Figure 5.8: $\Sigma(1/2)$.

Here is a hand-drawn portion of the surface $\Sigma(2/5)$. The complete picture would be a bit too complicated to see clearly on paper. The cell division has 256 vertices, 512 edges and 258 faces. (In particular, $\Sigma(2/5)$ is a sphere.) This picture is colored as the previous one. The outer square is not part of the surface. It is just the cutoff for the picture. The computer program lets you zoom in and out of the picture and see the whole thing.

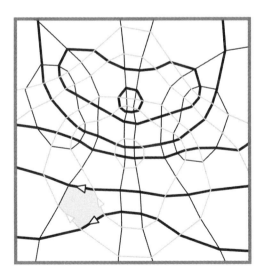

Figure 5.9: Part of $\Sigma(2/5)$.

Remark: In §5.6 below we will discuss the notion of a gear structure on the set of cycles in a plaid surface. The shaded polygon in Figure 5.9 will illustrate our definition.

5.6 DISCUSSION AND SPECULATION

Here I will discuss some topics for further study and engage in some speculation.

Topology: Here is a conjecture about the topology of the plaid surfaces.

Conjecture 5.2 *The following is true:*

1. *The genus of any plaid surface is divisible by 8.*

2. *Amongst all the plaid surfaces associated to the parameter p/q, the surface $\Sigma(p/q)$ has the maximum genus.*

3. *If $q_1 \leq q_2$ and p is relatively prime to both q_1 and q_2, then we have the inequality $G(p/q_1) \leq G(p/q_2)$.*

4. *$G(p/q) = 0$ if and only if $q \leq 5p + 1$.*

Conjecture 5.2 says in particular that all the p/q plaid surfaces are spheres if and only if $q \leq 5p + 1$. I checked Conjecture 5.2 for most parameters up to $p \leq 7$ and $q \leq 40$. After that, the computation gets too slow. I do not have the faintest idea how to prove the conjecture.

My program is not powerful enough to let me guess a general formula for the genus, but here are a few cases where I can guess with some confidence. Here is a chart of the genus $G(1/q)$ of $\Sigma(1/q)$ for some small values of q.

$$
\begin{array}{c|ccccccccccc}
q = & 2 & 4 & 6 & 8 & 10 & 12 & 14 & 16 & 18 & 20 & 22 \\
\frac{1}{8}G(1/q) = & 0 & 0 & 0 & 1 & 2 & 4 & 6 & 9 & 12 & 16 & 20
\end{array} \tag{5.2}
$$

The difference sequence is $0, 0, 0, 1, 2, 2, 3, 3, 4, 4, ...$ I could imagine giving a proof in this relatively trivial case, but I haven't.

Here is a chart of the genus $G(2/q)$ of $\Sigma(2/q)$ for some small values of q.

$$
\begin{array}{c|ccccccccccc}
q = & 3 & 5 & 7 & 9 & 11 & 13 & 15 & 17 & 19 & 21 & 23 \\
\frac{1}{8}G(2/q) = & 0 & 0 & 0 & 0 & 0 & 2 & 5 & 8 & 11 & 16 & 22
\end{array} \tag{5.3}
$$

Computing a bit further out, we see that the difference sequence seems to be $0, 0, 0, 0, 2, 3, 3, 3, 5, 6, 6, 6, 8, 9, 9, 9, 11, ...$ Again, I have no proof.

Based on scant evidence I would guess that the genus of $\Sigma(p_0/q)$ grows quadratically in q when p_0 is held fixed.

Gear Structure: After the proof of Lemma 5.1 we remarked that the oriented polygons in any integer XY slice (the plaid polygons) have a gear structure. Here we discuss further instances of gear structures.

In the next chapter we will give orientations to the XZ-cycles and the YZ-cycles. These orientations are canonical up to a global reversal of the orientations. We define the same notion of meshing as the gear structure remark. It is a consequence of our Curve Turning Theorem from the next chapter that these orientations impart gear structures to the polygons within each XZ and YZ slice. See Figure 6.1 for an example. (The gear structure does not exist on the quotient pictures we usually show; we have to use the oriented model.)

We want to say that our orientations *also* give rise to gear structures on the plaid surfaces. We need a definition of meshing first. Referring to the 3-colored 1-skeleton of the tiling on the plaid surface Σ, say that two same-colored cycles C_1 and C_2 *mesh* if they have edges e_1 and e_2 that belong to the same tile of Σ and have just one other edge between them along the boundary of the tile. If e_1 and e_2 are oriented the opposite way around the tile we say that these edges are *geared*. We say that C_1 and C_2 are geared if every pair of meshing edges is geared. (There might be multiple pairs of meshing edges.) The lightly shaded tile in Figure 5.9 shows what we mean.

The gear structure for the XY-cycles follows from the fact that every instance of a particle has the same transverse direction. Here is a sketch for the XZ direction. If e_1 and e_2 are two meshed XZ edges separated by an XY edge then they are geared thanks to the orientation part of Theorem 1.4. If e_1 and e_2 are separated by a YZ edge, then the kind of conservation principle discussed in the gear structure remark ought to imply that e_1 and e_2 are geared. The same argument should work in the YZ case, too.

Hidden Symmetry: Let $\widehat{\Omega}$ be the 3D cubical version of the unoriented plaid model. Again, $\widehat{\Omega}$ is an orbifold made by identifying the top and bottom XY faces of a cube and mirroring the other faces. As we remarked above, the YZ slices of $\widehat{\Omega}$ have order 8 dihedral symmetry. In particular, each such slice admits an order 4 rotation which interchanges the Y and Z directions.

This seems too good to be true. One might wonder why doing this rotation simultaneously in every YZ slice doesn't lead to a symmetry of $\widehat{\Omega}$ which interchanges the Y and Z directions. This would be a contradiction because the polygons in the XY slices, the plaid polygons, are not really like the polygons in the XZ slices.

The way out of the bind is to remember that when we draw pictures like Figures 5.1-5.3, we are vertically translating the picture so that the center of symmetry lies in the center of the figure. In $\widehat{\Omega}$, these centers of symmetry vary from slice to slice. It appear that they vary linearly up to a small additive error. Thus, if we string together all the order 4 rotations, we should get a locally affine map $A : \widehat{\Omega} \to \widehat{\Omega}$ which preserves each plaid surface up to a small additive error. This is to say that the surface $A(\Sigma)$ can be isotoped back to Σ by an isotopy that moves points by just a few units.

A carries the XZ slices to some non-coordinate slices which contain polygons that are essentially equivalent to the ones in the XZ slices. We assign the same capacity to a slice $A(\Pi)$ as we assign to Π. Thus, we now have 3 independent directions – XZ, YZ, and $A(XZ)$ – having these slices that look like pixelated spacetime diagrams (and also Truchet tilings).

Special Slices: The plaid surfaces seem to have a special relationship to the slices of capacity $2p$. These are the slices that arise in the Truchet Comparison Theorem from §7.3. See also Lemma 4.6. Here is something I noticed experimentally.

Conjecture 5.3 *Each plaid surface intersects each capacity $2p$ slice in at most one loop.*

For complicated rational parameters – i.e., those with long continued fractions – there are many different loops in these particular slices, having many different sizes. This follows from P. Hooper's Renormalization Theorem, [**H**, Theorem 11], which we discuss very briefly in §7.4, and our Truchet Comparison Theorem. Thus, if the above conjecture is true, it says that there are many plaid surfaces in $\widehat{\Omega}$ when we have a complicated rational parameter.

Comparison of Parameters: In §3.5, we showed how it was sometimes possible to get a good correspondence between the large-scale features of the plaid model in the 0th block B_0 with respect to two different parameters p_1/q_1 and p_2/q_2. This works well when p_1/q_1 and p_2/q_2 are close and also τ_1/ω_1 and τ_2/ω_2 are close.

Let $\widehat{\Omega}_j$ be the domain for the plaid surfaces with respect to p_j/q_j. We think of $\widehat{\Omega}_j$ as a cube of side length ω_j, even though technically $\widehat{\Omega}_j$ is better thought of as a certain orbifold whose interior agrees with the open cube of side length ω_j. Again, the blocks with respect to p_j/q_j are the integer XY slices of $\widehat{\Omega}_j$.

To make the picture clearer, we rescale so as to identify $\widehat{\Omega}_1$ and $\widehat{\Omega}_2$ with the unit cube. We call the rescaled versions $R\widehat{\Omega}_1$ and $R\widehat{\Omega}_2$. Once we rescale, the two 0th blocks are the bottoms (and the tops) of the cubes. We suggest to match up a block at height t in $R\widehat{\Omega}_1$ with the block in $R\widehat{\Omega}_2$ that has height as close as possible to t. When the parameters have large denominators, the two heights involved are quite close.

The same heuristic as in §3.5 suggests that (with the above hypotheses on the parameters) the typical slice at one parameter looks quite close to the typical slice at another parameter, provided that we only look at fairly low capacity lines. Indeed, there should be a close resemblance between the large-scale features of the plaid surfaces with respect to p_1/q_1 and the plaid surfaces with respect to p_2/q_2.

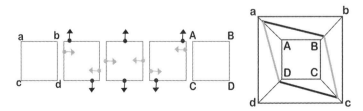

Figure 5.10: Ambiguous rectangles turn into an unambiguous solid.

Let's explain how the added dimension helps address the problem of ambiguous rectangles. The left half of Figure 5.10 suggests a portion of the fundamental cycle in which an ambiguous rectangle persists for a certain time and is flanked on either (time-)end by rectangles with no decoration. The right half of Figure 5.10 shows the corresponding rectangular solid in the 3D model. The letters help explain how the individual slices are stacked up.

While it is impossible to figure out the connectors in the individual slices on the left, one can fill in the entire solid with a kind of saddle surface which stays fairly close to the portion of the plaid surface that intersects this cube. In short, with the added dimension, the approximation process discussed in the previous chapter seems to have a more robust 3D analogue.

Chapter Six

Pixellation and Curve Turning

6.1 CHAPTER OVERVIEW

In this chapter we will revisit the idea of pixelated spacetime diagrams considered in the previous chapter. We will prove a technical result, the Curve Turning Theorem, which gives us a way to understand the spacetime diagrams in terms of patterns of oriented lines. In the next chapter we will use the Curve Turning Theorem to prove the Truchet Correspondence Theorem, the result which connects the plaid model to Truchet tilings.

The pixelation process looks like it is very delicate. If we jiggle the lines a bit, perhaps we get a completely different pattern. The curve turning process is quite robust, and the Curve Turning Theorem makes it clear that the pixelated spacetime diagrams are also quite robust. Figure 7.4 in the next chapter shows a good example of this phenomenon.

In §6.2 we will assign directions to all the particle lines. To do this, we use a key feature of the oriented plaid model which we have already established: the directions of all instances of a particle are the same. After we assign these directions, we define a process in which we follow each line along its direction and turn at each intersection. This process turns out to be closely related to the Truchet tile system that we will consider in the next chapter. We state the Curve Turning Theorem at the end of this section.

In §6.3 we prove a technical result about the spacing of the particle lines in our spacetime diagrams. This result will help with the proof of the Curve Turning Theorem. In §6.4 and §6.5 we prove the Curve Turning Theorem respectively in the vertical and the horizontal case. The result says that the curve turning process leads to essentially the same paths as the pixelation process defined in the previous chapter.

In §6.6 we give two applications of the Curve Turning Theorem.

6.2 ORIENTING THE WORLDLINES

We now explain how to assign orientations to the worldlines in a spacetime diagram. We direct a horizontal worldline – meaning a worldline corresponding to a horizontal particle – upward (respectively downward) if the transverse orientation points upward (respectively downward). We direct a vertical worldline upward (respectively downward) if its transverse direction points to the right (respectively left). The choice of up/down versus left/right is not important. Were we to pick a different convention, all the orientations within a diagram would be reversed and we would get the same overall result. The left side of Figure 6.1 shows an example of a vertical capacity 4 spacetime diagram for the parameter 2/5. The right-hand side shows the corresponding pixelated spacetime diagram.

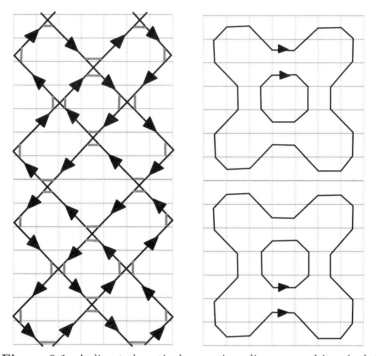

Figure 6.1: A directed vertical spacetime diagram and its pixelation.

Exercise: Take your finger and follow the lines on the left-hand side of Figure 6.1, according to the orientations. Turn at every intersection point, again following the orientations. The little grey segments serve as guides for your finger. You will see that the paths you trace out are essentially the same as the ones on the right-hand side of Figure 6.1. This fact, suitably generalized, is the content of the Curve Turning Theorem.

To describe the curve turning process in general, we choose some extremely small $\epsilon > 0$. We follow along the paths of the directed worldlines and then turn at every intersection. To get embedded paths, we don't go all the way to the intersections. Rather we stop when we arrive ϵ units before the intersection point, then cut across to the point on the next directed worldline which is ϵ units after the intersection point. The light grey segments on the left-hand side of Figure 6.1 show these shortcuts. We call this operation the *curve turning process*. The choice of ϵ is not important; two sufficiently small choices of ϵ would produce equivalent sets of curves in the sense we now define.

Equivalence: We call a piecewise linear homeomorphism $\phi : \Sigma \to \Sigma$ *small* if it moves no point by more than 1 unit. We say that two collections Γ and Γ' are *equivalent* if there is a small piecewise linear homeomorphism ϕ which maps Γ to Γ'. We use piecewise linear homeomorphisms because they automatically map polygons to polygons. However, the fact that the homeomorphisms are piecewise linear is not really an essential feature of the main result.

Theorem 6.1 (Curve Turning) *For any even rational parameter p/q and any capacity $\kappa \in \{2, 4, ..., 2p\}$, let Σ be any of the corresponding directed spacetime diagrams of capacity κ. Then the pixelation process and the curve turning process produce equivalent sets of polygons.*

We prove this result in the next three sections.

6.3 THE SPARSENESS OF WORLDLINES

Here we prove some technical results which will help with the proof of the Curve Turning Theorem. We work with the oriented plaid model.

Lemma 6.2 *Let L_1 and L_2 be two slanting lines having the same type, and mass less than $2p$. Let y_1 and y_2 be the Y-intercepts of L_1 and L_2 respectively. If $|y_1 - y_2| = 1$ then L_1 and L_2 must have opposite signs. Furthermore, it is impossible that $|y_1 - y_2| = 2$.*

Proof: Suppose first that $|y_1 - y_2| = 1$. The signed masses are given by the representatives of $2py_1$ and $2py_2$ mod 2ω which lie in $(-\omega, \omega)$. The difference between these two representatives is congruent to $2p$ mod 2ω. But it is impossible to have two integers of the same sign in $(-2p, 2p)$ whose difference is congruent to $2p$ mod 2ω. Hence L_1 and L_2 have opposite signs.

Suppose now that $|y_1 - y_2| = 2$. If this situation does happen, then we can find 2 numbers in $(-2p, 2p)$ whose difference is congruent to $4p$ mod 2ω. But $4p < 2p + 2q < 2\omega$. But then there is no number in $(-4p, 4p)$ that is congruent to $4p$ mod 2ω. $\square$

Lemma 6.3 *Let Σ be a slice of 3D plaid model having capacity at most $2p$. Two parallel worldlines cannot intersect the same unit integer square in Σ. In particular, the pixelated spacetime diagram in Σ has at most one connector in each tile.*

Proof: Figure 6.2 shows the situation in several cases. In all cases, the vertical spacing between these worldlines is the same as the distance between the Y-intercepts of the slanting lines of the same type through the many instances of the two corresponding particles in the same block.

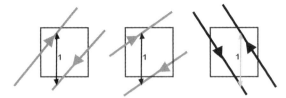

Figure 6.2: Two parallel worldlines going through the same tile.

Since $\max(1, P, Q) < 2$, the vertical spacing between the two worldlines is, in all cases, less than 3. Since the vertical spacing between the slanting lines is an integer, the only possiblities for the vertical spacing are 1 and 2. The latter case would imply the existence of two slanting lines of the same type and mass less than $2p$ whose Y-intercepts are 2 units apart. This contradicts Lemma 6.2. Hence, the vertical spacing must be 1 in all cases.

The slanting lines both give rise to light points on the same vertical grid line of capacity at most $2p$, so they have the same sign. But then we have found two slanting lines of the same sign and type, whose Y-intercepts differ by 1. This contradicts the first statement of Lemma 6.2. $\square$

6.4 CURVE TURNING THEOREM: VERTICAL CASE

Suppose that ζ_1 and ζ_2 are vertical light points on the same vertical grid line. We define the $s(\zeta_1, \zeta_2)$ to be the (integer) distance between the midpoints of the unit integer segments that contain ζ_1 and ζ_2.

Lemma 6.4 *Suppose that ζ_1 and ζ_2 are vertical light points, on the same vertical grid line, having different types. Then $s(\zeta_1, \zeta_2)$ is even if and only if ζ_1 and ζ_2 have the opposite transverse directions.*

Proof: Suppose ζ_1 lies on a slanting line L_1 of slope $-P$ and ζ_2 lies on a slanting line L_2 of slope $-Q$. Let ζ_1' be the reflection of ζ_1 in the horizontal midline of the relevant block. By symmetry, ζ_1' and ζ_1 have opposite transverse directions and $s(\zeta_1, \zeta_1')$ is even. So, the parity of $s(\zeta_1', \zeta_2)$ is the same as that of $s(\zeta_1, \zeta_2)$. At the same time, the transverse direction of ζ_1' is opposite that of ζ_1.

Let L_1' be the slanting line of slope $-Q$ through ζ_1'. Note that $s(\zeta_1', \zeta_2)$ is just the distance between the Y-intercepts of L_1' and L_2. Note that L_1' and L_2 have the same sign because they both contain light points on the same vertical grid line. Hence, by the argument in Lemma 1.3, their directions match if and only if the difference between their Y-intercepts is even. So, the transverse directions of ζ_1' and ζ_2 are the same if and only if $s(\zeta_1', \zeta_2)$ is even. This means that the transverse directions of ζ_1 and ζ_2 are different if and only if $s(\zeta_1, \zeta_2)$ is even. $\square$

Let Σ be a spacetime diagram for a vertical particle. By hypotheses, Σ has capacity at most $2p$.

Lemma 6.3 says that there can be at most 2 worldlines intersecting a tile. When there is just one worldline, we define a homeomorphism in T so that it fixes the vertices of T and moves the endpoints of $L \cap T$ to the corresponding midpoints of the edges of T. Figure 6.3 shows a canonical way to define the homeomorphism in a representative case. The homeomorphism maps each triangle on the left to the corresponding triangle on the right in an affine way.

Figure 6.3: A small piecewise linear homeomorphism in a tile.

Suppose there are two worldlines intersecting a tile. Figure 6.4 shows a complete set of representative cases modulo reflections. The shaded figures on the left-hand sides lead to matching pictures for the pixelation process and the unshaded figures on the right lead to mismatching pictures. The black points on the right side correspond to light points which violate Lemma 6.4.

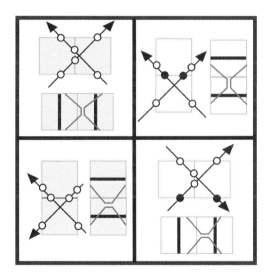

Figure 6.4: Two worldlines intersecting.

Let's explain the problem. In the figure on the top right, the two black points correspond to vertical light points ζ_1 and ζ_2 having different types and the same direction and $s(\zeta_1, \zeta_2) = 0$. In the figure on the bottom right, the two black points correspond to vertical light points having different types and opposite directions and $s(\zeta_1, \zeta_2) = 1$.

Figure 6.5 suggests how to construct a small piecewise linear homeomorphism ϕ between the relevant portions of Σ. We have added some shading to help the reader see how the homeomorphism works. Again, it is affine on each triangle.

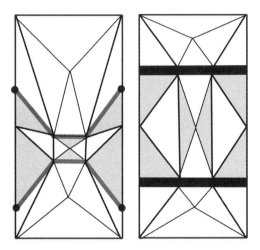

Figure 6.5: A small piecewise linear homeomorphism in a union of 2 tiles.

The union of two tiles is a 2×1 rectangle R. Our map ϕ preserves the two midpoints of the long sides of ∂R. Hence, ϕ preserves each unit integer segment in ∂R. On the unit integer segments of ∂R which have no connectors, ϕ acts as the identity. On the unit integer segments of ∂R which have connectors, ϕ acts in the most obvious way – the same way that the homeomorphism acts in the one worldline case considered first. Hence, all the individual homeomorphisms piece together continuously to give a small homeomorphism of Σ which carries the one set of polygons to the other. $\square$

6.5 CURVE TURNING THEOREM: HORIZONTAL CASE

We first prove an analogue of a special case of Lemma 6.4.

Lemma 6.5 *Two horizontal light points of different types and the same transverse directions cannot lie on the same horizontal integer segment.*

Proof: Let ζ_1 and ζ_2 be the two light points. Suppose that ζ_1 has type P and ζ_2 has type Q. Now ζ_1 lies on a slanting line of slope $\pm P$ and ζ_2 lies on a slanting line of slope $\pm Q$. Let $(m + 1/2, y)$ denote the midpoint of the horizontal unit segment containing ζ_1 and ζ_2. By symmetry it suffices to consider the case when $m \geq 0$. Let d_1 and d_2 be the distance from ζ_1 and ζ_2 respectively to this midpoint. Note that $d_1, d_2 \in [0, 1/2]$. Moreover, $\min(d_1, d_2) < 1/2$. Since $P + Q = 2$, the 4 quantities $\pm P d_1 \pm Q d_2$ all lie in $(-1, 1)$.

Let y_1 denote the Y-intercept of the slanting line L_1 of slope $-P$ which contains ζ_1. Let y_2 denote the Y-intercept of the slanting line L_2 of slope Q which contains ζ_2. Since $P + Q = 2$, we have

$$|y_1 - y_2| = (P + Q)(m + 1/2) \pm P d_1 \pm Q d_2 = 2m + 1 \pm P d_1 \pm P d_1 = 2m + 1.$$
$$(6.1)$$

The justification for the last equality is that the left-hand side of the equation is an integer and the last term lies in $(-1, 1)$. We have also shown, incidentally, that the last term in Equation 6.1 is 0.

Since ζ_1 and ζ_2 are both light points on the same horizontal line, L_1 and L_2 have the same sign. But then, since the difference between their Y-intercepts is odd, they have different directions. Hence ζ_1 and ζ_2 have opposite transverse directions. $\square$

Let Σ be a slice of capacity at most $2p$ corresponding to the horizontal case. The analysis here is just like in the vertical case. We just explain the differences. Figure 6.6 shows the analogue of Figure 6.4 in the horizontal case. There are a few more possibilities in the horizontal case, but the remaining possibilities are essentially the same as the ones shown.

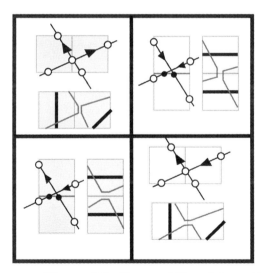

Figure 6.6: Two worldlines intersecting.

Using Lemma 6.5 in place of Lemma 6.4 we rule out the top right case just as we did in the vertical case. Consider the bottom right case. We will use notation consistent with the proof of Lemma 4.5. Let ζ_1^* and ζ_2^* denote the left endpoints contained on the two worldlines. (These are the points on the left edge of the spacetime diagram lying in the two worldlines.) Let $m \in 2\mathbf{Z}$ be the distance between these points. Let ζ_j be the light point in the planar model corresponding to ζ_j^*. Let y_j be the Y-intercept of the slanting line of slope $-P$ through ζ_j. In the even case, Equation 4.11 tells us that $y_1 - y_2$ is even. But then our two light points have the same direction. (See Lemma 1.3.) This contradicts the bottom right figure in Figure 6.6.

The rest of the proof is just like in the vertical case. We omit the description of the explicit local homeomorphisms.

6.6 TWO APPLICATIONS

Here we deduce a corollary about the surface $\Sigma(p/q)$ discussed in the previous chapter. $\Sigma(p/q)$ is the surface that contains the big polygon from Theorem 0.7.

Lemma 6.6 *For any even rational parameter p/q, the surface $\Sigma(p/q)$ is the only plaid surface that intersects the capacity 2 slices. In the unoriented model, the intersection of $\Sigma(p/q)$ with any capacity 2 slice is a single polygon.*

Proof: Let Π be a capacity 2 slice in the 3D version of the plaid model. The corresponding oriented spacetime diagram contains two directed polygonal paths. Within each block, the corresponding light points correspond to a single oriented plaid polygon and hence have opposite transverse directions. So, combinatorially speaking, and up to a global reversal of the directions, the picture in Π must look like Figure 6.7.

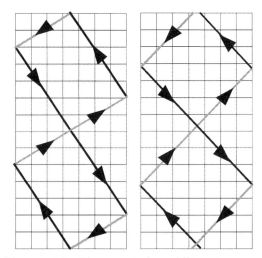

Figure 6.7: The oriented worldlines in capacity 2 slices.

In the oriented model, the curve turning process yields two loops, just as in Figure 4.5. When we pass to the unoriented plaid model, we get just one loop. In short, the union of plaid surfaces intersects Π in a single loop. Since the different plaid surfaces are disjoint, and a loop is connected, we see that there can be just one plaid surface intersecting Π.

Since the polygon from Theorem 0.7 intersects at least one vertical grid line of capacity 2 and both horizontal grid lines of capacity 2, we see that $\Sigma(p/q)$ intersects at least one vertical slice of capacity 2 and both horizontal slices of capacity 2. From the rotational symmetry discussed in §2.2, there must be a plaid surface Σ' which intersects the other vertical slice of capacity 2 and both horizontal slices of capacity 2. But then both these surfaces intersect the same horizontal slice(s) of capacity 2. Hence $\Sigma' = \Sigma(p/q)$. This shows that $\Sigma(p/q)$ intersects all 4 capacity 2 slices. In all cases, the intersection is a single loop. $\square$

Now we consider what happens for the capacity 4 slices. We pick p/q with $p \geq 2$ so that the capacity 4 slices satisfy the Curve Turning Theorem. For ease of exposition we just work with the vertical diagrams. As we discuss at the end of §7.6, the picture for the horizontal diagrams is isotopic to the picture for the vertical diagrams. The reader should look again at Figure 5.2 for an illustration of this result.

Lemma 6.7 *In the unoriented 3D model, the union of the plaid surfaces intersects the vertical slices of capacity 4 in two nested loops. The two loops are both oriented the same way, either clockwise or counterclockwise.*

Proof: Figure 6.8 shows the two possible arrangements for the worldlines in the diagram of half of a vertical capacity 4 slice. Up to a global reversal of the orientations, these are the only possibilities. The first arrangement leads to two loops in the undirected model, as claimed, and the second arrangement leads to four loops in the undirected model.

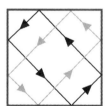 

Figure 6.8: The oriented worldlines in capacity 2 slices.

Note that reflection in the horizontal line $y = \omega/2$ is a symmetry of the undirected plaid model which reverses directions in the directed model. Given our definition of orientations of the world lines, this symmetry is compatible with the figure on the left in Figure 6.8 and incompatible with the figure on the right. Hence, only the figure on the left occurs. $\square$

Chapter Seven

Connection to the Truchet Tile System

7.1 CHAPTER OVERVIEW

We fix some even rational parameter p/q as usual. The main purpose of this chapter is to show that the pixelated spacetime slices of capacity $2p$ are combinatorially equivalent to certain of the tilings from P. Hooper's Truchet tile system [**H**].

In §7.2 we will describe the Truchet tile system. Our description is somewhat different from what is done in [**H**] but it is entirely equivalent.

In §7.3 we state our main result, the Truchet Comparison Theorem. One can view the Truchet Comparison Theorem as a computational tool for understanding some of the pixelated spacetime diagrams.

In §7.4 we use the Truchet Comparison Theorem to get more information about the surface $\Sigma(p/q)$ from Corollary 6.6.

In §7.5 we prove a curious result from elementary number theory which underlies the Truchet Comparison Theorem. After a modest amount of searching, I was unable to locate this result in the literature.

In §7.6 we put together the ingredients and prove the Truchet Comparison Theorem.

7.2 TRUCHET TILINGS

In this section we describe the Truchet tilings in [**H**] with $\alpha = \beta$, where $\alpha = p'/q' \in (0, 1/2]$. We insist that p' is odd and q' is even. We will see that every such tiling appears infinitely often as a pixelated spacetime diagram. Figure 7.1 shows the two Truchet tiles having "slope" $+1$ and -1.

Figure 7.1: The truchet tiles.

The tilings in [**H**] corresponding to the parameters α, β are described at the beginning of §3 of [**H**]. In general, the definition of the tiling depends on some choice of offset $(x, y) \in \mathbf{R}^2$, but when α and β are rational, all offsets give the same tiling up to translation. We take $\alpha = \beta$ and $x = y = -\alpha/2$. The functions ω and η in [**H**] are the same function.

Following [**H**], but using our notation, we define $\eta : \mathbf{Z} \to \{-1, 1\}$ to be the sign of

$$(m\alpha - \alpha/2)_1. \tag{7.1}$$

In other words, we take the sign of the representative of $m\alpha - \alpha/2 \mod 1$ in $(-1/2, 1/2)$. This quantity is never 0, and the sequence is periodic with period q'. We call this sequence the *Truchet sequence*. We call the finite subsequence $\eta_1, ..., \eta_{q'}$ the *first period*.

Here are some examples.

- When $\alpha = 1/2$, the sequence $\{\eta_m\}$ is given by $+-$ repeating.

- When $\alpha = 1/4$ the sequence is $+ + --$ repeating.

- When $\alpha = 3/8$ the sequence is $+ - - + - + + -$ repeating.

We define a tiling of the plane in which the Truchet tiles are centered at integer points (m, n) and the slope of the Truchet tile is the same sign as $\eta_m \eta_n$. Figure 7.2 shows one period of the periodic Truchet tiling produced by these parameters.

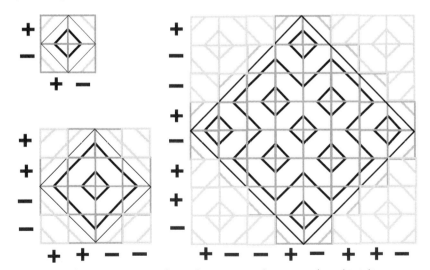

Figure 7.2: Truchet tile patterns for $\alpha = 1/2, 1/4, 3/8$.

The sequences above all have the form $(A)(-A)(A)(-A), ...$ where A is a palindrome. The first period is given by $(A)(-A)$. The next result shows that this is always true.

Lemma 7.1 *For any parameter, the first period of the Truchet sequence has the form $(+A)(-A)$ where A is a palindrome.*

Proof: Let $\alpha = p/q$. Let $f(m) = m\alpha - \alpha/2$. The mth term in the Truchet sequence is given as $\tau(m) = (f(m))_1$. The first period runs for $m = 1, ..., q$. We compute that

$$f(m) + f(q + 1 - m) = p \in \mathbf{Z}, \qquad f(q/2 + m) - f(m) = p/2 \in \frac{1}{2}\mathbf{Z} - \mathbf{Z}.$$
$$(7.2)$$

The first equation implies that $\tau(m)$ and $\tau(q + 1 - m)$ have opposite signs. The second equation implies that $\tau(m)$ and $\tau(q/2 + m)$ have opposite signs. Putting these two facts together gives us the lemma. $\square$

Given the structure of our sequence, we can always isolate a diamond shaped fundamental domain whose boundary is an impenetrable barrier disjoint from all Truchet paths. We call this barrier the *diamond*.

7.3 THE TRUCHET COMPARISON THEOREM

The Truchet tiles are centered at integer lattice points in $\mathbf{R}^2$. The horizontal and vertical grid lines go through the centers of these tiles. We can give directions to the integer grid lines using the Truchet tiling. We direct the vertical line $x = m$ upward if $\eta_m = 1$ and downward if $\eta_m = -1$. We direct the horizontal line $y = n$ rightward if $\eta_n = 1$ and leftward if $\eta_n = -1$. When we do the curve turning process we get a family of curves that is equivalent to the family of Truchet polygons. This is a purely local result, and easy to see from a single example. This fact is the basis for our main result below.

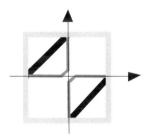

Figure 7.3: Curve following for the Truchet tiles.

Let $\Gamma : \mathbf{R} \to \mathbf{R}$ denote the infinite dihedral group generated by the maps $x \to x - 1$ and $x \to 1 - x$. A fundamental domain for the action of Γ is the interval $[0, 1/2)$. Given any $x \in \mathbf{R}$, let $D(x)$ denote the representative of Γx in the fundamental domain. The map D plays an important role in [**H**], as we will see below.

Given a parameter p/q we define

$$\alpha(p/q) = D(P^{-1}) = D\left(\frac{p + q}{2p}\right). \qquad (7.3)$$

For instance,

$$\alpha(4/9) = D(13/8) = 3/8, \qquad \alpha(4/15) = D(19/8) = 3/8.$$

We chose this example because we have already drawn the Truchet pattern for $\alpha = 3/8$. Figure 7.2 shows half of the pixelated spacetime diagram for the vertical slice of capacity $8 = 2 \times 4$. Figures 7.2 and 7.4 illustrate our main result. Notice the similarity between the Truchet pattern for

$$3/8 = \alpha(4/9) = \alpha(4/15)$$

and the capacity 8 spacetime diagrams for 4/9 and 4/15.

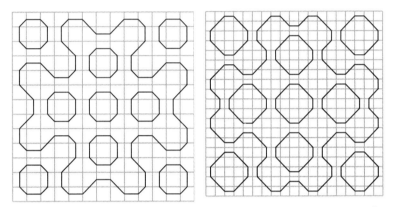

Figure 7.4: The vertical capacity 8 diagrams the parameters 4/9 and 4/15.

Before we state our main result, we need one more notion. Within a Truchet tiling, we say that two curves *kiss* if they intersect the same Truchet tile. Within a pixelated spacetime diagram, we way that two polygons *kiss* if the corresponding orbits of the strict curve turning process meet at a vertex. The strict process traces over the curves exactly. Here is our main result.

Theorem 7.2 (Truchet Comparison) *Let $p/q \in (0,1)$ be an even rational parameter. Then the union of curves in the pixelated spacetime diagram of capacity $2p$ for the parameter p/q is isotopic to the union of curves in two consecutive fundamental domains for the Truchet tiling associated to the parameter $\alpha(p/q)$. The isotopy preserves the kissing relation.*

Computer Tie-In: Here is how to see the Truchet Comparison Theorem in action for the vertical spacetime diagrams. (You can configure the program in a similar way to see the theorem for the horizontal spacetime diagrams.) Open up the main program. When the main program is running, open up the *spacetime diagrams* window, the *truchet* window, and the *spacetime diagram* control panel. Turn off all the features except *background*, *grid*, and *pixelated V*. Select a random parameter p/q. Beneath the spacetime control panel, set the *frame* option to p and the capacity option to $2p$. Select the *V* option at the bottom. Click on the *spacetime diagrams* window and then enter the r key. This will recenter the picture.

7.4 THE FUNDAMENTAL SURFACE

Here we extract more information about the plaid surface from Corollary 6.6. Our information derives from Hooper's Renormalization Theorem about the above Truchet tilings, namely [**H**, Theorem 11]. The Renormalization Theorem from [**H**] is rather involved and we just want a simple corollary. We will state the corollary and sketch how to get it from the Renormalization Theorem.

Notice the big plus-shaped polygon on the right side of Figure 7.2. This polygon kisses its 4 neighbors across all 4 sides of the diamond barrier. The following lemma shows that this always occurs.

Lemma 7.3 (Big Loop) *In any of the Truchet tilings there is an infinite, doubly periodic family of polygons such that adjacent polygons in the family kiss each other across the diamond barriers.*

Proof: (Sketch) We introduce the map $R : [0, 1/2) \to [0, 1/2]$.

$$R(\alpha) = D\left(\frac{\alpha}{1 - 2\alpha}\right). \tag{7.4}$$

Here D is the map from the Truchet Comparison Theorem. For instance

$$R(13/34) = 3/8, \qquad R(3/8) = 1/2.$$

R is not defined on $1/2$. R decreases the denominator, so for any rational odd-over-even parameter $\alpha \in (0, 1/2]$ there is some n such that $R^n(\alpha) = 1/2$.

We proceed by induction on the size of the denominator. Call a Truchet polygon *small* if it just involves 4 tiles. Otherwise call it *big*. Let T denote the infinite Truchet tiling associated to α and let T' denote the infinite Truchet tiling associated to $\alpha' = R(\alpha)$. The Renormalization Theorem says (among other things) that there is a bijection between the big polygons in T and all the polygons in T', and this bijection respects the symmetries of the tilings and preserves the kissing relations. Thus, if T' has a family of polygons that kiss each other across adjacent diamond fundamental domains, so does T. Since the tiling associated to $\alpha = 1/2$ has such a family of polygons, all the tilings do. $\square$

We now combine the Big Loop Lemma with the Truchet Comparison Theorem. Recall that $\Sigma(p/q)$ is the plaid surface that contains the plaid polygon from Theorem 0.7. We work with the unoriented plaid model, so that our 3D domain Ω is the cube $[0, \omega]^3$ with suitable identifications and mirrorings on its boundary. See §5.2.

Theorem 7.4 (Filling) *For every even rational paramter p/q, the plaid surface $\Sigma(p/q)$ intersects every integer XZ and YZ slice of nonzero capacity in $[0, \omega]^3$ and every integer XY slice except perhaps one.*

Proof: Choose some YZ slice Π. We will study the vertical pixelated spacetime diagram in Π. Let $\widehat{\Pi}$ denote the lift of the pixelated spacetime diagram to $\boldsymbol{R}^2$ under the (orbifold) universal covering map we considered in the proof of the Truchet Comparison Theorem. We call the polygons in $\widehat{\Pi}$ *spacetime loops*, to distinguish them from the Truchet polygons.

The proof of the Truchet Comparison Theorem works on the level of the universal cover, and shows that there is a bijection between the spacetime loops in Π and the Truchet polygons that preserves the kissing relations. In particular, there is a family of spacetime loops which corresponds to the family of polygons from the Big Loop Lemma. Each member in this family kisses the 4 adjacent spacetime loops. Hence, there is a spacetime loop $\gamma \subset \Pi$ which intersects every vertical integer segment of Π except the vertical boundaries of Π.

Let Σ' be the plaid surface containing γ. One of the vertical segments just discussed is contained in a XZ slice of capacity 2. Hence γ intersects an XZ slice of capacity 2. But then Σ' intersects an XZ slice of capacity 2. We have already proved in Theorem 6.6 that only $\Sigma(p/q)$ intersects such slices. Hence $\Sigma' = \Sigma(p/q)$. In short, we have shown that $\Sigma(p/q)$ intersects Π. Since Π is arbitrary, $\Sigma(p/q)$ intersects every integer YZ slice of nonzero capacity.

The same argument works in the horizontal case. Hence $\Sigma(p/q)$ intersects every integer XZ slice of nonzero capacity.

Finally, the spacetime loop γ also intersects every horizontal integer segment except the ones at the top and bottom of Π. Hence γ intersects every XY slice of Ω except for one. Hence Σ intersects every XY slice of Ω except perhaps for one. $\square$

7.5 A RESULT FROM ELEMENTARY NUMBER THEORY

We give the proof of the Truchet Comparison Theorem in the section following this one. Let p/q be an even rational parameter. As usual, let $\omega = p + q$ and $\widehat{\tau}$ be such that $2p\widehat{\tau} \equiv 1 \bmod \omega$. We will carry along an example to help explain the result. In the example, $p/q = 5/8$. This gives $\omega = 13$ and $\widehat{\tau} = 4$. Consider the following sequences.

Red Sequence: Define $A_k = \pm 1$ according to the sign of the representative of $(\omega/2 - \omega k) \bmod 2p$ that lies in $(-p, p)$. This is the red sequence. In the example, the sequence of representatives is $(3/2, 9/2, -5/2, 1/2, 7/2)$ and the red sequence is $+ + - + +$. One can see that this sequence is related to the definition of the Truchet tilings.

Blue Sequence: Take the numbers $\widehat{\tau}, 3\widehat{\tau}, 5\widehat{\tau}, ..., (2p-1)\widehat{\tau} \bmod \omega$ and sort them so that they appear in order. Call the resulting sequence $\{y_k\}$. We define $B_k = +1$ if y_k is even and $B_k = -1$ if y_k is odd. The blue sequence is $\{B_k\}$. In the example, the unsorted sequence is $4, 12, 7, 2, 10$. The sorted sequence is $2, 4, 7, 10, 12$. The blue sequence is $+ + - + +$. One can see from Lemma 2.3 that this sequence is related to the signed masses of the slanting lines in the plaid model.

Lemma 7.5 *The red and blue sequences coincide.*

Proof: Let $\{y_k\}$ be as in the definition of the blue sequence. Let $\langle m \rangle_p$ denote the representative of m mod p that lies in $\{1, ..., p\}$. Let $\langle m \rangle_{2p}$ denote the representative of m mod $2p$ in $\{-p + 1, ..., p\}$.

Let $c_1, ..., c_p \in \{1, ..., p\}$ be such that

$$y_k \equiv \widehat{\tau}(2c_k - 1) \bmod \omega, \qquad k = 1, ..., p. \tag{7.5}$$

We will derive the following formula:

$$c_k = py_k + \frac{\omega + 1}{2} - \omega k. \tag{7.6}$$

If we declare that the sign of 0 is -1 then A_k is always the sign of

$$\widehat{c}_k = \left\langle \frac{\omega + 1}{2} - k\omega \right\rangle_{2p}. \tag{7.7}$$

From Equation 7.6 and the fact that $c_k \in \{1, ..., p\}$ we see that when y_k is even (respectively odd) we have $\widehat{c}_k = c_k$ (respectively $\widehat{c}_k = c_k - p$). Hence y_k is even if and only if $A_k = +1$. □

To finish the proof of Lemma 7.5 we establish Equation 7.6. When $p = 1$ we have $c_1 = 1$ and $y_1 = (\omega + 1)/2$. So, the formula holds. Henceforth assume that $p \geq 2$. We set $y_0 = 0$ and $c_0 = 0$ for notational convenience. Define

$$\delta_k = c_{k+1} - c_k, \qquad \Delta_k = y_{k+1} - y_k. \tag{7.8}$$

Lemma 7.6 $|\Delta_i - \Delta_j| \leq 1$ *for all* i, j.

Proof: Suppose $\Delta_i \leq \Delta_j - 2$. Let

$$c'_{j+1} = c_j + \delta_i + \epsilon p.$$

Here $\epsilon \in \{-1, 0, 1\}$ is chosen so that $c'_{j+1} \in \{1, ..., p\}$. This is always possible because all the c variables are in $\{1, ..., p\}$. Since $2\widehat{\tau}p \equiv 1 \bmod \omega$,

$$y'_{j+1} \equiv \widehat{\tau}(2c'_{j+1} - 1) \equiv \widehat{\tau}(2c_j - 1) + \Delta_i + \epsilon = y_j + \Delta_i + \epsilon \quad \bmod \omega.$$

But then $y_j \leq y'_{j+1} < y_{j+1}$. Since y_j and y_{j+1} are consecutive, $y'_{j+1} = y_j$. But then $\Delta_i = 1$, which means that $2\widehat{\tau}\delta_i \equiv 1 \bmod \omega$. But this forces $\delta_i \equiv p$ mod ω. If $i > 0$ this is impossible because $c_i \neq c_{i+1} \in \{1, ..., p\}$. If $\Delta_0 = 1$ then $y_1 = 1$. But then $\widehat{\tau}(2c_1 - 1) \equiv \widehat{\tau}(2p) \bmod \omega$ for some $c_1 \in \{1, ..., p\}$. This is also impossible. □

Lemma 7.7 $\omega/2 < p\Delta_k < 3\omega/2$ *for all* $k = 0, ..., p - 1$.

Proof: It is convenient to set $y_{p+1} = \omega + 1$. Observe that for each y_i there is some other y_j such that $y_i + y_j = \omega + 1$. But then $\Delta_p = \Delta_0$. We also have

$$\omega + 1 = y_{p+1} = \sum_{k=0}^{p} \Delta_k. \tag{7.9}$$

Let $\underline{\Delta}$ be the min of Δ_j and let $\overline{\Delta}$ be the max. If $\underline{\Delta} = \overline{\Delta}$ then the common value is $(\omega + 1)/(p + 1)$ and the inequalities are obvious.

Otherwise, we have $\overline{\Delta} = \underline{\Delta} + 1$, by the previous result. For the upper bound, we have $\underline{\Delta} \leq \omega/(p + 1)$, and this gives

$$p\overline{\Delta} = p\underline{\Delta} + p < \frac{p}{p+1}(\omega) + p < 3\omega/2.$$

For the lower bound we have $\overline{\Delta} \geq (\omega + 2)/(p + 1)$, and this gives

$$p\underline{\Delta} \geq \frac{p}{p+1}(\omega + 2) - p = \frac{p(q+1)}{p+1} > \omega/2.$$

A bit of calculus establishes the last inequality for $1 < p < q$. $\square$

Let c_k^* denote the right hand side of Equation 7.6. We want to show that $c_k^* = c_k$. Using $2\hat{\tau} \equiv 1 \mod \omega$, we compute that $c_k^* \equiv c_k \mod \omega$. Also, Lemma 7.7 tells us that

$$c_1^* = y_1 + \frac{\omega + 1}{2} - \omega = \Delta_0 + \frac{\omega + 1}{2} - \omega \in (0, \omega).$$

Since $c_1 \in (0, \omega)$ as well, we must have $c_1 = c_1^*$.

Suppose by induction that $c_k = c_k^*$. Then

$$c_{k+1}^* - c_{k+1} = (c_{k+1}^* - c_k^*) - (c_{k+1} - c_k) = (p\Delta_k - \omega) - \delta_k \in (-\omega, \omega).$$

The last entry of this equation comes from the fact that $|\delta_k| \leq p < \omega/2$, and from Lemma 7.7. Since $c_{k+1}^* \equiv c_{k+1} \mod \omega$ and since $|c_{k+1}^* - c_{k+1}| < \omega$, we must have $c_{k+1}^* = c_{k+1}$. This completes the induction step. This establishes Equation 7.6.

7.6 PROOF OF THE TRUCHET COMPARISON THEOREM

We will give the proof for the spacetime diagrams corresponding to the slices of positive sign. The other case follows from symmetry.

There are $2p$ integers $0 < y_1 < ... < y_{2p} < 2\omega$ such that the slanting lines having these y-intercepts have positive sign and mass less than $2p$. We call this the *low mass sequence*.

Lemma 7.8 *The sequence of directions of the low mass sequence coincides with the first period of the Truchet sequence, up to a global sign.*

Proof: This is an application of the fact that the red and blue associated sequences coincide. Let $\alpha = \alpha(p/q)$. Let $(+A)(-A)$ be the first period of the Truchet sequence associated to α. Let $\{A_k\}$ be the first half of this first period.

Let $\{y_k\}$ be the first half of the low mass sequence. Thanks to Lemma 2.3, this sequence is exactly the one used to define the red sequence above. All the slanting lines in question have the same sign, and so their directions are determined by their parities. (See Lemma 1.3.) Therefore, up to a global sign, the sequence of directions of the first half of the low mass sequence is the red sequence associated to the parameter. To finish the proof we just have to recognize the first half of the first period of the Truchet sequence as the blue sequence associated to the parameter. The fact that the second halves agree follows from symmetry.

Let $[x]_s$ denote the sign of $(x)_s$. Let $\{A_k\}$ denote the first half of the first period of the Truchet sequence. The sequence associated to $\alpha + 1$ is $(+A)(-A)$ and the sequence associated to $1 - \alpha$ is $(-A)(+A)$. Hence, if $D(\alpha) = D(\alpha')$ then the associated Truchet sequences agree up to a global sign. Since $D(P^{-1}) = \alpha$, we see that

$$A_k = [P^{-1}k - P^{-1}/2]_1.$$

Multiplying through by $2p$ we get $\pm A_k = [\omega k - \omega/2]_{2p}$. The sign out in front is a global one. Hence the first half of the first period of the Truchet sequence agrees with the blue sequence. $\square$

Consider the vertical case of the Truchet Comparison Theorem. To compare the pixelated spacetime diagrams to the Truchet tilings, we first extend the spacetime diagrams so that they fill the whole plane. We track the particles not just in the fundamental blocks but in the entire plane. This is to say that we consider the light points not just on the segment of a vertical grid line contained in a block, but on the entire line, and also we repeat the fundamental period endlessly. By symmetry, this extended picture is simply the orbifold universal cover of the original diagram. That is, we extend in the vertical direction by translations and in the horizontal direction by reflections. Figure 7.5 shows how we have reflected the original spacetime diagram in Figure 5.1 across one of its mirrored boundaries. Note how the directions match.

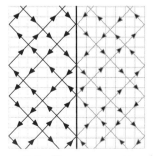

Figure 7.5: Two consecutive fundamental domains.

Once we make the extension, the worldlines naturally fit together to make an infinite grid of directed lines of slope ± 1. The lines of slope $+1$ intersect any given horizontal line in our diagram in the same positions that the slanting lines of slope $-P$ intersect the corresponding vertical grid line. The lines of slope -1 intersect the same horizontal line at the points where the slanting lines of slope $-Q$ intersect the corresponding vertical lines.

Note that the slanting lines of a given type intersect any vertical line in the same order as their y-intercepts. Indeed, the pattern of intersection in any vertical line is the same, up to translation, as the pattern of intersection on the y-axis. Therefore, by Lemma 7.8, the sequence of directions associated to the lines of slope $+1$ in the extended spacetime diagram is the same as the Truchet sequence, up to a global sign change. The same goes for the lines of slope -1.

Given that the Truchet tiling can also be described in terms of curve turning, we see that the extended pixelated spacetime diagram is isotopic to the Truchet tiling in a way which preserves the kissing relation. The only difference is that we have turned the picture 45 degrees and also adjusted the spacing between the lines. (Unlike in the Truchet tiling, the spacing is not always the same between different pairs of lines in the spacetime diagram.) Moreover, the original spacetime diagram contains 2 consecutive fundamental domains for the extended tiling. This completes the proof in the vertical case but below we will further analyze the symmetry of the situation.

The proof in the horizontal case is essentially the same. We just describe the differences. The first difference is that the lines in the extended spacetime model have slopes P and $-Q$ rather than 1 and -1. This doesn't make any difference from a combinatorial point of view. Second, while the order in which the slanting lines intersect a given horizontal line is the same as the y-intercepts, the two sets are not translation equivalent. They are related by an affine transformation. This does not make any difference. This completes the proof of the Truchet Comparison Theorem

Remarks: Our argument incidentally proves that the horizontal and vertical pixelated spacetime diagrams of capacity κ are combinatorially isomorphic to each other for all $\kappa \leq 2p$. The point is that both diagrams come from the curve turning construction, and in both slices we see the same combinatorial structure to the directions of the lines involved.

Part 2. The Plaid PET

Chapter Eight

The Plaid Master Picture Theorem

8.1 CHAPTER OVERVIEW

This chapter starts Part 2 of the monograph. The purpose of this chapter is to prove Theorem 1.4 and Theorem 0.3, the Plaid Master Picture Theorem. We will deduce both of these results from Theorem 8.2 below. Subsequent chapters in this part are devoted to proving Theorem 8.2. This part of the monograph only depends on §1, §2.2, §2.3, §4.2, §4.3, §4.4, and §5.2.

Here is an abstract description of Theorem 8.2. Let A be an even rational parameter and let $P = 2A/(1+A)$. Let $\mathcal{T}$ denote the set of possible directed connectors which join midpoints of edges in the unit square. Let Π denote the grid of unit integer squares in the plane. A *tiling classifying space* is a pair (Φ_A, X_P), where $\Phi_A : \Pi \to X_P$ is a map and X_P is a space partitioned into subsets, each of which is labeled by an element of $\mathcal{T}$. To each unit integer square $\square$ we assign a connector based on which partition set of X_P contains the image $\Phi_A(c)$, where c is the center of $\square$.

Theorem 8.2 says that the union PL_A of plaid polygons is generated by an explicitly defined tiling classifying space (Φ_A, X_P). Moreover, there is a nice space X which has the individual spaces X_P as rational slices. The space X has a partition into convex polytopes, and one obtains the partition of X_P by intersecting the relevant slice with this partition.

In §8.2 we describe the space X. In §8.3 we will describe the partition of X into convex integer polytopes. We call the partition the *checkerboard partition*. In §8.4 we will explain the classifying map $\Phi_A : \Pi \to X_P$. In §8.5 we state Theorem 8.2 and deduce Theorem 1.4 and Theorem 0.3 from it.

Computer Tie-In: Here is how to see Theorem 8.2 in action. In the main program, open the *Plaid PET* popup window. On the *plaid PET slicer* window, set the *PET choice* to *neutral*. Drag the mouse over the *plaid PET slicer* to explore the space X and the checkerboard partition. (See the *document* window for more operating instructions.) Next, open the *planar* window and the *plaid* control panel. Click on the *oriented tiles* option on the *plaid* control panel and satisfy yourself that the oriented tiles you see match the plaid polygons. Next, click on an oriented tile and observe the point $\Phi_A(c) \in X_P$ appear in the *plaid PET* window. (The window will show the slice X_P.) Here $c \in \Pi$ is the center of the tile you clicked and A is the selected parameter and $P = 2A/(1+A)$. The tiles and the polyhedra in the *plaid PET* window are colored to highlight the classifying map.

8.2 THE SPACES

Define

$$\widehat{X} = \boldsymbol{R}^3 \times [0,1]. \qquad (8.1)$$

The coordinates on $\widehat{X}$ are given by (x, z, y, P). We think of

$$P = \frac{2A}{1+A}, \qquad A = p/q, \qquad (8.2)$$

but P is allowed to take on any real value in $[0,1]$. Define the following affine transformations of $\widehat{X}$.

- $T_1(x, y, z, P) = (x + 2, y + P, z + P, P)$.

- $T_2(x, y, z, P) = (x, y + 2, z, P)$.

- $T_3(x, y, z, P) = (x, y, z + 2, P)$.

Define two abelian groups of affine transformations:

$$\Lambda' = \langle T_1, T_2, T_3 \rangle, \qquad \Lambda = \langle T_1^2, T_2, T_3 \rangle. \qquad (8.3)$$

Finally define

$$X' = \widehat{X}/\Lambda', \qquad X = \widehat{X}/\Lambda. \qquad (8.4)$$

Both spaces should be considered flat affine manifolds - i.e., manifolds whose overlap functions are restrictions of affine transformations. All the affine transformations in sight preserve the slices $\boldsymbol{R}^3 \times \{P\}$ and act as translations on these slices. Thus X' and X are fibered by 3-dimensional Euclidean tori. We denote these tori by X'_P and X_P respectively. We associate these spaces respectively to the unoriented and oriented plaid models.

8.3 THE CHECKERBOARD PARTITION

Here we describe a partition of the space X' into convex integral polytopes. Taking the double cover, we get a partition of X. Lifting to $\widehat{X}$, we get a Λ'-invariant partition of $\widehat{X}$ into infinitely many polytopes. When we add labels, this infinite partition is only Λ-invariant.

The rectangular solid

$$[-1, 1]^3 \times [0, 1] \qquad (8.5)$$

serves as a fundamental domain for the action of Λ' on $\widehat{X}$. We think of this space as a fiber bundle over the (x, P) plane. The base space B is the rectangle $[-1, 1] \times [0, 1]$. The space B has a partition into 3 triangles, as shown in Figure 8.1.

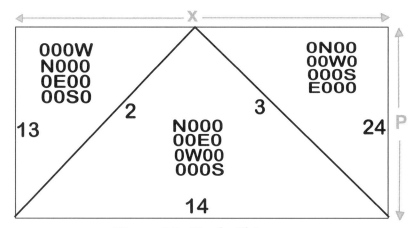

Figure 8.1: The (x, P) base space.

Each triangular region has been assigned a 4×4 matrix which we will explain momentarily. The slices above each open triangle in the partition of B consists of a 4×4 grid of rectangles. Four of the rectangles are squares. These correspond to polytopes labeled by the empty set. (In Figure 8.2 we replace the empty set label by one of the symbols in $\{N, E, W, S\}$ because this makes the rest of the labeling pattern more clear.) The pattern of the squares is indicated by the nonzero entries in the matrices. For instance, the grid shown in Figure 8.2 corresponds to a fiber over the left triangle. We have also indicated the labels in the picture.

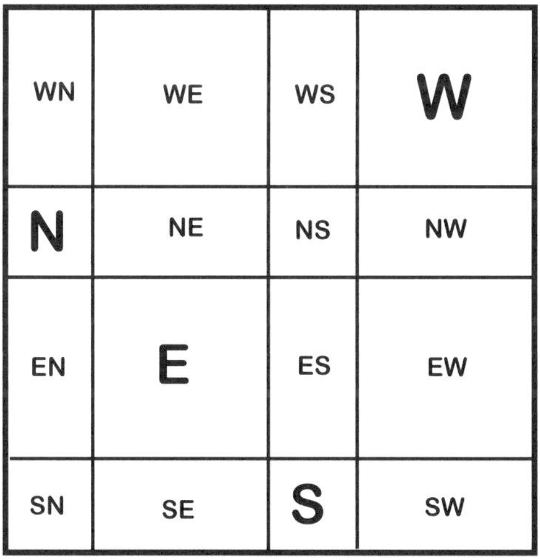

WN	WE	WS	**W**
N	NE	NS	NW
EN	**E**	ES	EW
SN	SE	**S**	SW

Figure 8.2: The checkerboard partition.

The labels can be derived from the matrices in Figure 8.1. Every non-square rectangle in a column has the same first label and every rectangle in a row has the same second label.

The edge labels in Figure 8.1 encode degenerations in the fibers. Away from the singular edges of the base space, each polytope is a rectangle bundle over an open triangle. As one approaches the edges of the triangles in the base, the partition degenerates in that some of the rectangles shrink to line segments or points. The numbers on the edges in Figure 8.1 indicate which rows and columns degenerate. For instance, as we approach the edge labeled 2, the thickness of the second column from the left tends to 0 and at the same time the thickness of the second row from the bottom tends to 0. The (unlabeled) picture is symmetric with respect to reflection in the line $y = x$.

From the description of these degenerations and from the fact that we know we are looking at slices of convex integral polytopes, one can actually reconstruct the entire partition from the labelings in Figure 8.1. Each polytope in Figure 8.2 comes with a label. For instance, the label of the polytope which intersects the fiber in the northwest corner is $\{W, N\}$.

Thus, each polytope in the partition of $\widehat{X}$ is either labeled by the $\emptyset$ symbol, or else by a pair $\alpha \neq \beta$, where $\alpha, \beta \in \{N, E, W, S\}$. Now we explain how we enhance the labeling so that each polytope not labeled by $\emptyset$ is labeled by an *ordered pair* (α, β). Our idea is to explicitly give the ordering on a list of 10 polytopes (below) and then use the action of Λ' to propagate the ordering to the whole partition. The group Λ preserves the ordered labeling and elements of $\Lambda' - \Lambda$ reverse the ordering. We found the ordered labeling (which turns out to be canonical) by trial and error.

Here we list the vertices and ordered labels of 10 polytopes. All the polytopes in the partition of $\widehat{X}$ can be obtained from these using the group generated by Λ' and the following two additional elements:

- **Negation:** The map $(x, y, z, P) \to (-x, -y, -z, P)$ preserves the partition and changes the labels as follows: N and S are swapped and E and W are swapped.

- **Flipping:** The map $(x, y, z, P) \to (x, z, y, P)$ preserves the partition and changes the labels as follows: N and S are swapped, and the order of the label is reversed.

The last polytope listed has 12 vertices, and the listing is spread over 2 lines.

$$
\begin{bmatrix} -1 \\ -1 \\ -1 \\ 0 \end{bmatrix}
\begin{bmatrix} -1 \\ +1 \\ -1 \\ 0 \end{bmatrix}
\begin{bmatrix} +1 \\ +1 \\ -1 \\ 0 \end{bmatrix}
\begin{bmatrix} +1 \\ +1 \\ +1 \\ 0 \end{bmatrix}
\begin{bmatrix} 0 \\ 0 \\ 0 \\ +1 \end{bmatrix}
\qquad (W, E)
$$

$$
\begin{bmatrix} +1 \\ +1 \\ -1 \\ 0 \end{bmatrix}
\begin{bmatrix} +1 \\ +1 \\ 0 \\ +1 \end{bmatrix}
\begin{bmatrix} +1 \\ +1 \\ -1 \\ +1 \end{bmatrix}
\begin{bmatrix} 0 \\ +1 \\ -1 \\ +1 \end{bmatrix}
\begin{bmatrix} 0 \\ 0 \\ -1 \\ +1 \end{bmatrix}
\qquad (E, S)
$$

$$
\begin{bmatrix} +1 \\ -1 \\ -1 \\ 0 \end{bmatrix}
\begin{bmatrix} +1 \\ 0 \\ 0 \\ +1 \end{bmatrix}
\begin{bmatrix} +1 \\ 0 \\ -1 \\ +1 \end{bmatrix}
\begin{bmatrix} 0 \\ 0 \\ -1 \\ +1 \end{bmatrix}
\begin{bmatrix} 0 \\ -1 \\ -1 \\ +1 \end{bmatrix}
\qquad (E, N)
$$

$$\begin{bmatrix} -1 \\ +1 \\ -1 \\ 0 \end{bmatrix} \begin{bmatrix} +1 \\ +1 \\ -1 \\ 0 \end{bmatrix} \begin{bmatrix} +1 \\ +1 \\ +1 \\ 0 \end{bmatrix} \begin{bmatrix} 0 \\ 0 \\ 0 \\ +1 \end{bmatrix} \begin{bmatrix} 0 \\ +1 \\ 0 \\ +1 \end{bmatrix} \begin{bmatrix} +1 \\ +1 \\ 0 \\ +1 \end{bmatrix} \begin{bmatrix} +1 \\ +1 \\ +1 \\ +1 \end{bmatrix} \qquad (W, S)$$

$$\begin{bmatrix} -1 \\ -1 \\ -1 \\ 0 \end{bmatrix} \begin{bmatrix} +1 \\ -1 \\ -1 \\ 0 \end{bmatrix} \begin{bmatrix} +1 \\ +1 \\ -1 \\ 0 \end{bmatrix} \begin{bmatrix} 0 \\ 0 \\ -1 \\ +1 \end{bmatrix} \begin{bmatrix} 0 \\ 0 \\ 0 \\ +1 \end{bmatrix} \begin{bmatrix} +1 \\ 0 \\ 0 \\ +1 \end{bmatrix} \begin{bmatrix} +1 \\ +1 \\ 0 \\ +1 \end{bmatrix} \qquad (S, W)$$

$$\begin{bmatrix} -1 \\ +1 \\ +1 \\ 0 \end{bmatrix} \begin{bmatrix} -1 \\ 0 \\ 0 \\ +1 \end{bmatrix} \begin{bmatrix} -1 \\ 0 \\ +1 \\ +1 \end{bmatrix} \begin{bmatrix} -1 \\ +1 \\ 0 \\ +1 \end{bmatrix} \begin{bmatrix} -1 \\ +1 \\ +1 \\ +1 \end{bmatrix} \begin{bmatrix} 0 \\ +1 \\ +1 \\ +1 \end{bmatrix} \begin{bmatrix} -2 \\ 0 \\ 0 \\ +1 \end{bmatrix} \qquad \emptyset$$

$$\begin{bmatrix} -1 \\ -1 \\ -1 \\ 0 \end{bmatrix} \begin{bmatrix} +1 \\ -1 \\ -1 \\ 0 \end{bmatrix} \begin{bmatrix} -1 \\ -1 \\ -1 \\ +1 \end{bmatrix} \begin{bmatrix} 0 \\ 0 \\ 0 \\ +1 \end{bmatrix} \begin{bmatrix} 0 \\ 0 \\ -1 \\ +1 \end{bmatrix} \begin{bmatrix} 0 \\ -1 \\ 0 \\ +1 \end{bmatrix} \begin{bmatrix} 0 \\ -1 \\ -1 \\ +1 \end{bmatrix} \begin{bmatrix} +1 \\ 0 \\ 0 \\ +1 \end{bmatrix} \qquad (S, N)$$

$$\begin{bmatrix} +1 \\ +1 \\ -1 \\ 0 \end{bmatrix} \begin{bmatrix} +1 \\ -1 \\ -1 \\ 0 \end{bmatrix} \begin{bmatrix} +1 \\ +1 \\ 0 \\ +1 \end{bmatrix} \begin{bmatrix} +1 \\ 0 \\ 0 \\ +1 \end{bmatrix} \begin{bmatrix} +1 \\ +1 \\ -1 \\ +1 \end{bmatrix} \begin{bmatrix} +1 \\ 0 \\ -1 \\ +1 \end{bmatrix} \begin{bmatrix} 0 \\ 0 \\ -1 \\ +1 \end{bmatrix} \begin{bmatrix} 2 \\ +1 \\ 0 \\ +1 \end{bmatrix} \qquad (E, W)$$

$$\begin{bmatrix} -1 \\ -1 \\ +1 \\ 0 \end{bmatrix} \begin{bmatrix} -1 \\ -1 \\ 0 \\ +1 \end{bmatrix} \begin{bmatrix} 0 \\ -1 \\ 0 \\ +1 \end{bmatrix} \begin{bmatrix} 0 \\ 0 \\ 0 \\ +1 \end{bmatrix} \begin{bmatrix} 0 \\ -1 \\ +1 \\ +1 \end{bmatrix} \begin{bmatrix} 0 \\ 0 \\ +1 \\ +1 \end{bmatrix} \begin{bmatrix} +1 \\ 0 \\ +1 \\ +1 \end{bmatrix} \begin{bmatrix} +1 \\ -1 \\ +1 \\ 0 \end{bmatrix} \qquad \emptyset$$

$$\begin{bmatrix} -1 \\ -1 \\ -1 \\ 0 \end{bmatrix} \begin{bmatrix} -1 \\ +1 \\ +1 \\ 0 \end{bmatrix} \begin{bmatrix} -1 \\ -1 \\ +1 \\ 0 \end{bmatrix} \begin{bmatrix} -1 \\ +1 \\ -1 \\ 0 \end{bmatrix} \begin{bmatrix} +1 \\ +1 \\ +1 \\ 0 \end{bmatrix} \begin{bmatrix} -1 \\ -1 \\ -1 \\ +1 \end{bmatrix}$$

$$\begin{bmatrix} -1 \\ 0 \\ -1 \\ +1 \end{bmatrix} \begin{bmatrix} -1 \\ -1 \\ 0 \\ +1 \end{bmatrix} \begin{bmatrix} -1 \\ 0 \\ 0 \\ +1 \end{bmatrix} \begin{bmatrix} 0 \\ 0 \\ 0 \\ +1 \end{bmatrix} \begin{bmatrix} -3 \\ -1 \\ -1 \\ 0 \end{bmatrix} \begin{bmatrix} -2 \\ -1 \\ -1 \\ +1 \end{bmatrix} \qquad \emptyset$$

Computer Tie-In: On the main program open up the *plaid PET* window and the *poly info* window. If you click the middle mouse button (or use key-x) on the polytopes in the partition, the *poly info* window will show their vertices and their labels.

8.4 THE CLASSIFYING MAP

The *plaid grid* is defined to be the set Π of centers of integer unit squares. For each parameter $A \in (0,1)$, there is an affine map $\Phi_A : \Pi \to \widehat{X}$, given by

$$\Phi_A(x,y) = x(2P, 2P, 2P, 0) + y(2, 0, 2P, 0) + (0, 0, 0, P). \quad P = \frac{2A}{1+A}. \quad (8.6)$$

In this section we fix A and we restrict the action of Λ' to $\mathbf{R}^3 \times \{P\}$ and we identify this slice with $\mathbf{R}^3$. Thus, we think of Λ' and Λ as lattices of translations acting on $\mathbf{R}^3$. We also set $\Phi = \Phi_A$ for notational convenience. Let L' and L be the symmetry lattices for the unoriented and oriented plaid model respectively.

Let $\omega = p + q$. (The parameter is $A = p/q$.) Let $\mathbf{Z}_0$ and $\mathbf{Z}_1$ respectively denote the set of even and odd integers, we define

$$X^{\#} = \{(a/\omega, b/\omega, c/\omega) | \, a \in \mathbf{Z}_1, \, b, c \in \mathbf{Z}_0\}. \quad (8.7)$$

Note that the lattice Λ' consists entirely of vectors in $(\mathbf{Z}_0/\omega)^3$. Hence Λ' preserves $X^{\#}$. The same goes for Λ, a sub-lattice of Λ'. Here are some counts:

- $X^{\#}/\Lambda'$ is a finite set of points with ω^3 members.

- $X^{\#}/\Lambda$ is a finite set of points with $2\omega^3$ members.

- Π/L' is also a finite set of points with ω^3 members.

- Π/L is also a finite set of points with $2\omega^3$ members.

Here is the main result in this section. It is meant to hold with respect to any rational parameter $A = p/q$.

Lemma 8.1 *The following is true.*

1. *$\Phi(L) \subset \Lambda$ and $\Phi(L') \subset \Lambda'$.*

2. *$\Phi(\Pi) \subset X^{\#}$.*

3. *Φ induces a bijection from Π/L to $X^{\#}/\Lambda$, and likewise a bijection between Π/L' and $X^{\#}/\Lambda'$.*

4. *$X^{\#}$ is disjoint from the walls of the partition of $\widehat{X}$.*

Statements 2 and 4 combine to say that the set $\Phi(\Pi)$ lies in the unions of the interior of the pieces of the partition. This will make our constructions below well defined. We prove Lemma 8.1 through a series of smaller steps. The reader who is keen to get to the main point of the chapter should skip the proof of Lemma 8.1 on the first pass. The main thing we need to know in what follows is that the image $\Phi(\Pi)$ does not hit the walls of our partition.

Proof of Statement 1: In the proof we ignore the 4th coordinate of the map Φ. We will prove this for Λ'. The proof for Λ is similar. Recall that $A = p/q$, and $\omega = p+q$, and that L' is generated by $(\omega^2, 0)$ and $(0, \omega)$. Since Φ is a linear map – at least when we forget the last coordinate – it suffices to prove that $\Phi(\omega^2, 0)$ and $\Phi(0, \omega)$ both belong to Λ'. We compute

$$\Phi(\omega^2, 0) = (2P\omega^2, 2P\omega^2, 2P\omega^2) =$$

$$P\omega^2(2, P, P) + (0, 2P\omega^2 - P^2\omega^2, 0) + (0, 0, 2P\omega^2 - P^2\omega^2).$$

Note that $P = 2p/\omega$, so that $P\omega^2$ and $P^2\omega^2$ are both even integers. Hence, the last two vectors above have the form $(0, k, 0)$ and $(0, 0, k)$ for some even integer k. All the vectors on the second line of our equation belong to Λ'.

We compute

$$\Phi(0, \omega) = (2\omega, 0, 2P\omega) = \omega(2, P, P) + (0, -\omega P, 0) + (0, 0, P\omega).$$

Again, the last two vectors have the form $(0, -k, 0)$ and $(0, 0, k)$ for some even integer k. So, all the vectors on the right-hand side belong to Λ'. $\square$

Proof of Statement 2: Each point in Π has the form $c = (m+1/2, n+1/2)$ for $m, n \in \mathbf{Z}$. We compute

$$\Phi(c) = \frac{1}{\omega}\left(4pm + 2p + (2n+1)\omega, 4pm + 2p, 4pm + 2p + 4pn + 2p\right).$$

The first of these coordinates is in $\mathbf{Z}_1/\omega$ and the second two are in $\mathbf{Z}_0/\omega$. $\square$

Proof of Statement 3: We prove this for Λ'. The proof for Λ is similar, and indeed follows from symmetry and the other case. We know already that $\Phi(c) \in X^\#$ for any $c \in \Pi$. By Statements 1 and 2, the map Φ induces a map from Π/L' to $X^\#/\Lambda'$. In view of the fact that both sets have the same number of points, we just have to show that the induced map is an injection.

Suppose that $\Phi(c_1) = \Phi(c_2)$ for points $c_1, c_2 \in \Pi$. We write

$$c_2 - c_1 = (m, n), \qquad m, n \in \mathbf{Z}. \tag{8.8}$$

We remind the reader that Λ' is generated by the vectors $(2, P, P)$, $(0, 2, 0)$ and $(0, 0, 2)$. Recalling that $P = 2p/\omega$, we have

$$\Phi(c_2) - \Phi(c_1) = (2mP, 2mP, 2mP) + (2n, 0, 2nP) =$$

$$\left(\frac{4pm}{\omega} + 2n, \frac{4pm}{\omega}, \frac{4pm}{\omega} + \frac{4pn}{\omega}\right) \equiv$$

$$\left(\frac{4pm}{\omega}, \frac{4pm}{\omega} - \frac{2pn}{\omega}, \frac{4pm}{\omega} + \frac{2pn}{\omega}\right) \quad \text{mod } \Lambda'.$$

The first coordinate must be an even integer. Since 4 and p are relatively prime to ω, we have $m = k\omega$ for some $k \in \mathbf{Z}$. This gives us

$$\Phi(c_2) - \Phi(c_1) \equiv \left(4pk, 4pk - \frac{2pn}{\omega}, 4pk + \frac{2pn}{\omega}\right) \equiv$$

$$\left(4pk, -\frac{2pn}{\omega}, \frac{2pn}{\omega}\right) \equiv \left(0, -\frac{2pn}{\omega} - \frac{4kp^2}{\omega}, \frac{2pn}{\omega} - \frac{4kp^2}{\omega}\right) \quad \text{mod } \Lambda'.$$

The second and third coordinates must be even integers. In particular, the sum of the second and third coordinates must be an even integer. This sum is $-8kp^2/\omega$. Since 8 and p are both relatively prime to ω, this means that $k \equiv 0 \bmod \omega$. But then $m \equiv 0 \bmod \omega^2$.

Since $k \equiv 0 \bmod \omega$, the number $4kp^2$ is also an even integer. Hence

$$\Phi(c_2) - \Phi(c_1) \equiv \left(0, -\frac{2pn}{\omega}, \frac{2pn}{\omega}\right) \quad \text{mod } \Lambda'.$$

These last coordinates must be even integers. This forces $n \equiv 0 \bmod \omega$. Remembering that $m \equiv 0 \bmod \omega^2$ we now see that $c_2 - c_1 \in L'$. $\square$

Proof of Statement 4: As usual, we forget the last coordinate and work in $\mathbf{R}^3$. Say that a *special plane* is a plane of the form $(x, *, *)$ with $x \in \mathbf{Z}_1/\omega$. By definition, $X^{\#}$ is contained in the union of special planes.

We check that each special plane intersects the walls of the partition in lines of the form $y = u$ and $z = u$. When $x = -1$, the values of u are in the set $\{-1, 1 + P, 1\}$, all of which belong to $\mathbf{Z}_1/\omega$. As the value of x changes by $2/\omega$, the offsets for the wall-fiber intersections change by $\pm 2/\omega$. Hence, we always have $u \in \mathbf{Z}_1/\omega$. But then $X^{\#}$ does not hit any of these lines.

The proof is done, but we want to elaborate on the picture developed in the proof of the last result. If we place a cube of side length $2/\omega$ around each point of $X^{\#}$, then these cubes tile $\mathbf{R}^3$. Moreover, these cubes intersect each special plane in an $\omega \times \omega$ grid. The image $\Phi(\Pi)$ intersects this grid at the centers of the squares whereas the walls of the partition intersect the grid in line segments extending the edges of the squares. $\square$

8.5 THE MAIN RESULT

We fix the parameter A and use the same notation conventions from the previous section. The partition of $\widehat{X}$ by labeled polytopes descends to the partition of X. Because $\Phi(\Pi)$ is disjoint from the walls of our partition, we can define a tiling of $\mathbf{R}^2$ as follows. Given $c \in \Pi$ let $L(c)$ be the label of the polytope in the partition that contains $\Phi(c)$. Let $\square_c$ be the unit integer square centered at c. If $L(c) = \emptyset$ we put nothing in $\square_c$. If $L(c) = (\alpha, \beta)$, we draw the connector in $\square_c$ pointing from the midpoint of the α edge to the midpoint of the β edge.

We call this tiling the *master tiling*. Given a unit square $\square$ and an edge E of $\square$ we say that the master tiling *involves* $(\square, E)$ if the connector in $\square$ uses the midpoint of E. In this case, we call $(\square, E)$ *out-pointing* if the connector points to E and *in-pointing* if the connector points from E. Here is our main result, which holds for every parameter.

Theorem 8.2 *For any unit integer square $\square$ and any edge E of $\square$, the number of light points assigned to E is odd if and only if the master tiling involves $(\square, E)$. In this case, $(\square, E)$ is in-pointing if and only if the transverse direction associated to the light point on E points into $\square$.*

Proof of Theorem 1.4: For each square $\square$, there are either 0 or 2 edges E such that the master tiling involves $(\square, E)$. But these involved edges are precisely the ones having one light point on them. Hence, there are 0 or 2 relevant light points on the boundary of $\square$. This proves the first half of Theorem 1.4. For the second half, suppose that $\square$ has 2 edges involved in the tiling. Then the construction above tells us to draw a connector from one of these edges to the other. The connector is in-pointing at one of the edges and out-pointing at the other. But then the corresponding transverse directions at the relevant light points have the same property. $\square$

Consistency: Suppose $\square_1$ and $\square_2$ are two adjacent squares sharing a common edge E. If $(\square_1, E)$ is involved in the master tiling, then E is a relevant edge for the plaid model: It has one light point on it. But then Theorem 8.2 tells us that $(\square_2, E)$ is also involved in the master tiling. Similarly, if the classifying map connector in $\square_1$ points into E then the classifying map connector in $\square_2$ points out of E.

The Plaid Master Picture Theorem: Consider the map $f : \Pi \to \Pi$ that comes from simply following the arrows on the tiles. Given $c_0 \in \Pi$, the new point $C_1 = f(c_0)$ is defined to be the center of the tile into which the tile at c_0 points. For instance, the tile centered at c_0 is NE, then $c_1 = c_0 + (1, 0)$. In case the tile centered at c_0 is empty, we have $c_1 = c_0$.

Thanks to Theorem 8.2, there is a corresponding map $F : \widehat{X} \to \widehat{X}$ such that

$$F \circ \Phi = \Phi \circ f. \tag{8.9}$$

The map F has the following definition:

- F is the identity on polytopes labeled by $\emptyset$.

- $F(x, y, z, P) = (x, y, z, P) + (2, 0, 2P, 0)$ on polytopes whose label ends in N.

- $F(x, y, z, P) = (x, y, z, P) - (2, 0, 2P, 0)$ on polytopes whose label ends in S.

- $F(x, y, z, P) = (x, y, z, P) + (P, P, P, 0)$ on polytopes whose label ends in E.

- $F(x, y, z, P) = (x, y, z, P) - (P, P, P, 0)$ on polytopes whose label ends in W.

We can also interpret F as a map on X because everything in sight is Λ invariant. With this interpretation, (X, F) is the fibered integral affine PET.

The second partition is obtained by simply applying F to all the pieces in the partition. The pieces in the second partition do not overlap because the whole process is invertible: We can simply follow the plaid polygons in the other direction, and this corresponds to the action of F^{-1} on the second partition. Thus, when we take Φ mod Λ, Theorem 8.2 implies the Plaid Master Picture Theorem from the introduction.

We have more to say about the pair of partitions described above. The first partition is specially adapted to F, and the second partition is somewhat less natural. This might seem to violate an obvious symmetry between F, which corresponds to following the curves around in one direction, and F^{-1}, which corresponds to following the curves around in the other. We might have made a similar construction for F^{-1} by using the first label of each polytope rather than the second. This would lead to a very natural first partition for F^{-1} and a somewhat less natural partition for F.

Computer Tie-In: Open up the *plaid PET* window on the main program. On the *plaid PET slicer* window set the *PET choice* either to *forward* or *backward*. This will show you the second partitions for F or F^{-1} respectively.

Chapter Nine

The Segment Lemma

9.1 CHAPTER OVERVIEW

The next 4 chapters are devoted to the proof of Theorem 8.2. This chapter contains the statement and proof of the Segment Lemma. The two chapters following this one respectively contain the statement and proof of the Vertical Lemma and the Horizontal Lemma. The final chapter in Part 2 puts these 3 technical lemmas together and finishes the proof of Theorem 8.2. We recommend skipping the proofs of the technical lemmas on the first pass. In other words, after reading the introductory sections of the next 3 chapters, just skip to §12 to see the logic of the overall proof.

We fix a parameter $A = p/q$ and set $P = 2A/(1 + A)$. We continue the notation from the previous chapter. Recall that Π is the plaid grid. We set $\Phi = \Phi_A$ and $X = X_P$. We drop the 4th coordinate of our spaces and maps, so as to work in $\mathbf{R}^3$. This allows us to reserve the notation X_V and X_H, in this chapter, for certain 2-dimensional slices of X.

A particle (as defined in §4) in the directed plaid model breaks up into two *half-particles* in which the relative motion of the subset goes in one direction across the block. Given a vertical (respectively horizontal) half-particle $\{z_j\}$ we let $\{c_j\} \subset \Pi$ be the corresponding sequence in Π such that the unit integer square centered on c_j contains z_j in its west (respectively south) boundary edge. We call $C = \{c_j\}$ a *tracker*. We are interested in $\Phi(C)$.

A tracker can have one of 4 types: vertical P, vertical Q, horizontal P, and horizontal Q, depending on the nature of the half-particle it tracks. We call two trackers *partners* if they track the two halves of the same particle.

Let Π_V (respectively Π_H) denote the subset of Π consisting of points that lie on horizontal (respectively vertical) midlines of the fundamental blocks in the plaid model. (This definition depends on the parameter.) We will prove below that every vertical (respectively horizontal) particle tracker C contains one point of Π_V (respectively Π_H). We call this point the *anchor* and denote it by C_0 in all cases. We will also show that mod Λ we have

$$\Phi(\Pi_V) \subset X_V = \{(x, y, y) | x, y \in \mathbf{R}^2\}, \tag{9.1}$$

$$\Phi(\Pi_H) \subset X_H = \{(\pm 1, y, z) | y, z \in \mathbf{R}^2\}. \tag{9.2}$$

We say that C is *aimed by* a vector W if every point of $\Phi(C)$ is equivalent mod Λ to a point on the line segment with endpoints $\Phi(C_0) \pm W$. Here is the main technical result in this chapter.

Lemma 9.1 (Segment) *Let $C = \{c_j\}$ be a tracker.*

1. *If C is vertical P, then $\Phi(C)$ is aimed by $(0, 0, 1)$.*

2. *If C is vertical Q, then $\Phi(C)$ is aimed by $(0, 1, 0)$.*

3. *If C is horizontal P, then $\Phi(C)$ is aimed by (P, P, P).*

4. *If C is horizontal Q, then $\Phi(C)$ is aimed by $(Q, 0, 0)$.*

Figure 9.1 illustrates the Segment Lemma. The top left picture shows the two images associated to vertical partners. The picture has been projected into the YZ plane. The two segments are actually in two different YZ slices. The dotted line is the projection of X_V.

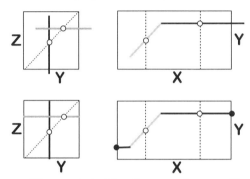

Figure 9.1: The Segment Lemma illustrated.

The top right picture shows the two images associated to horizontal partners. Here we are projecting into the XY plane. The projection into the XZ plane would look the same. This time the two images fit together as the picture suggests. The dotted line is the projection of X_H. In both cases, the white points are $\Phi(C_0)$ and differ from each other by $(2, P, P)$. Each bottom figure shows a Λ-equivalent set contained in the standard fundamental domain.

In §9.2 we explain the existence and image of the anchor point. In §9.3 we explain how to reinterpret the classifying map from §8.4 as a map defined on a certain subset of $\boldsymbol{R}^3$. Our technical result, Lemma 9.4, will be useful for the computations in this chapter. In §9.4 and 9.5 respectively we treat the vertical and horizontal cases of the Segment Lemma.

Computer Tie-In: Here is how to see the Segment Lemma in action. Open up the main program. Once the main program is running, open up the *Plaid PET* window and the *Plaid PET* control panel. Turn off the partition display. Now open the *sanity checks* control panel. If you select the *plaid V particle* option you can see a plot of the Φ image of a randomly chosen vertical particle. With this option on, every time you click on the *Plaid PET* window you get a new random instance. The *plaid H particle* option does the same for the horizontal particles. You can see finer plots by choosing p/q with both p and q large.

9.2 THE ANCHOR POINT

Recall that the anchor point of a tracker lies in X_V or X_H, depending on whether the tracker is vertical or horizontal.

Lemma 9.2 *Every tracker C has an anchor point.*

Proof: In the vertical case, the half-particles move (relatively speaking) up or down at speed one. Hence they occupy the central unit integer vertical segment. The corresponding point of the tracker is the anchor.

We treat the horizontal P case. The horizontal Q case has the same treatment. Consider a horizontal half-particle H of type P. There are $2P$ instances of H. The first instance of H lies on the west boundary of some block. The first instance of the type Q horizontal H' that is the continuation of H lies on the east boundary of some block. Since H moves linearly in a relative sense, there is an instance z of H that lies on the vertical midline of a block. The corresponding point of the tracker is the anchor. $\square$

Lemma 9.3 $\Phi(\Pi_V) \subset X_V$ *mod* Λ *and* $\Phi(\Pi_H) \subset X_H$ *mod* Λ.

Proof: Recall that Λ is generated by $(4, 2P, 2P)$ and $(0, 2, 0)$ and $(0, 0, 2)$. Any point of Π_V has the form $(x, \pm\omega/2)$. We have

$$\Phi(x, \pm\omega/2) = x(2P, 2P, 2P) + (\pm\omega, 0, \pm P\omega) \equiv (2Px \pm \omega, 2Px, 2Px) \text{ mod } \Lambda.$$

The equivalence comes from the fact that $P\omega = 2p$ is an even integer. This last point lies in X_V.

Any point of Π_H has the form $(k\omega/2, y)$ where k and $2y$ are odd integers. We have

$$\phi(k\omega/2, y) = (Pk\omega + 2y, *, *) = (2pk + 2y, *, *) \equiv (\pm 1, *, *) \text{ mod } \Lambda.$$

The last congruence comes from the fact that $2pk + 2p$ is an odd integer. We have not listed the second and third coordinates because we do not care about them for this calculation. The final point lies in X_H. $\square$

9.3 A COMPUTATIONAL TOOL

We use the stacking construction from §5.2 to help us compute the classifying map Θ. Recall that

$$\Omega = [0, \omega^2] \times [-\omega, \omega], \qquad \widehat{\Omega} = [0, \omega]^2 \times [0, 2\omega] \qquad (9.3)$$

are respectively the fundamental domain for the oriented plaid model and the domain for the 3D interpretation of the oriented plaid model.

Let $\widehat{\Pi}$ denote the set of points (x, y, z) with x and y half-integers and z an integer. As in Part 1 of the monograph, we identify the points c and $c + (0, \omega^2)$ to simplify our calculations. With this convention, we set

$$\widehat{c} = (x, y, z) \leftrightarrow (x, y) + (-z\widehat{\tau}, \epsilon(z)\omega) = c, \qquad z = 0, ..., 2\omega - 1. \qquad (9.4)$$

Here $\epsilon(z) = 0$ when z is even and $\epsilon(z) = -1$ when z is odd and $\widehat{\tau}$ is as in §1.2. This sets up a bijection between $\widehat{\Pi} \cap \widehat{\Omega}$ and $\Pi \cap \Omega$.

Lemma 9.4 *Suppose $\widehat{c}_j \leftrightarrow c_j$ for $j = 1, 2$. Then*

$$\Phi(c_2) - \Phi(c_1) \equiv \Phi_0(\widehat{c}_2 - \widehat{c}_1) \mod \Lambda,$$

where

$$\Phi_0(x, y, z) = x(2P, 2P, 2P) + y(2, 0, 2P) + z(-2, 0, 0). \qquad (9.5)$$

Proof: Given that Φ_0 is linear and Φ is an affine map into the torus $\mathbf{R}^3/\Lambda$, it suffices to check our formula on the standard basis of $\mathbf{R}^3$. If $\widehat{c}_2 - \widehat{c}_1 = (1, 0, 0)$ then $c_2 - c_1 = (1, 0)$ and the formula follows immediately from the definition of Φ. The same goes when $\widehat{c}_2 - \widehat{c}_1 = (0, 1, 0)$.

Suppose $\widehat{c}_2 - \widehat{c}_1 = (0, 0, 1)$. We have $c_2 - c_1 = (-\widehat{\tau}\omega, \pm\omega)$.

$$\Phi(-\widehat{\tau}\omega, 0) = (-2P\widehat{\tau}\omega, -2P\widehat{\tau}\omega, -2P\widehat{\tau}\omega) = (-4p\widehat{\tau}, -4p\widehat{\tau}, -4p\widehat{\tau}) \equiv$$

$$(-4p\widehat{\tau}, 0, 0) \equiv (0, 2Pp\widehat{\tau}, 2Pp\widehat{\tau}) \equiv (0, P, P) \mod \Lambda. \qquad (9.6)$$

The last equivalence comes from the fact that $2p\widehat{\tau} - 1 = K\omega$ for some integer K. This gives us

$$2Pp\widehat{\tau} - P = P(2p\widehat{\tau} - 1) = PK\omega = 2pK \in 2\mathbf{Z}.$$

Next, we compute

$$\Phi(0, \omega) = (2\omega, 0, 2P\omega) \equiv (2\omega, 0, 0) =$$

$$(2(\omega - 1) + 2, 0, 0) \equiv (2, -P\omega + P, -P\omega + P) \equiv (2, P, P) \mod \Lambda. \qquad (9.7)$$

What makes this work is that $2(\omega - 1)$ is divisible by 4. Equations 9.6 and 9.7 combine to give

$$\Phi(-\widehat{\tau}, \pm\omega) \equiv (0, P, P) \pm (2, P, P) \equiv (-2, 0, 0) \mod \Lambda. \qquad (9.8)$$

The last equation works for either sign choice because $(-2, 0, 0) \equiv (2, 2P, 2P)$ mod Λ. Putting all this together, we get Equation 9.5. $\square$

Here is an alternate formula for Φ_0. Recall that $Q = 2 - P$.

Lemma 9.5 *Suppose $x, y, z \in \mathbf{Z}$. Then*

$$\Phi_0(x, y, z) \equiv x(-2Q, 0, 0) + y(2, 0, -2Q) + z(-2, 0, 0) \mod \Lambda. \qquad (9.9)$$

Proof: We have

$$(-2Q, 0, 0) \equiv (2P - 4, 0, 0) \equiv (2P, 2P, 2P) \mod \Lambda,$$

$$(2, 0, -2Q) \equiv (2, 0, 2P) \mod \Lambda.$$

The result follows almost immediately. $\square$

9.4 THE VERTICAL CASE

Now we turn to the proof of the Segment Lemma in the vertical case.

Lemma 9.6 *Let C be a vertical P tracker. There is some vector W such that mod Λ the image $\Phi(C)$ lies in a segment with endpoints $W \pm (0,0,1)$.*

Proof: Let $C = \{c_k\}$. Suppose that c_k and c_{k+1} are the centers of 2 successive instances of the corresponding half-particle. Let $\widehat{c}_k$ and $\widehat{c}_{k+1}$ be the corresponding points in $\widehat{\Pi}$. We have

$$c_{k+1} - c_k = (-\widehat{\tau}\omega, 1), \qquad \widehat{c}_{k+1} - \widehat{c}_k = (0,1,1).$$

The first equation implies the second. By Lemma 9.4, we have

$$\Phi(c_{k+1}) - \Phi(c_k) \equiv \Phi_0(0,1,1) \equiv$$

$$(2,0,2P) + (-2,0,0) \equiv (0,0,2P) \bmod \Lambda. \qquad (9.10)$$

No matter what the value of t, we have $(0,0,t) \equiv (0,0,t') \bmod \Lambda$ for some $t' \in [0,2]$. Hence, mod Λ, all points of $\Phi(C)$ lie on the same segment having length 2 and parallel to $(0,0,1)$. $\square$

Lemma 9.7 *Let C be a vertical Q tracker. There is some vector W such that mod Λ the image $\Phi(C)$ lies in a segment with endpoints $W \pm (0,1,0)$.*

Proof: This time we have $\widehat{c}_{k+1} - \widehat{c}_k = (0,-1,1)$. This leads to

$$\Phi(c_{k+1}) - \Phi(c_k) \equiv \Phi_0(0,-1,1) \equiv$$

$$-(2,0,2P) + (-2,0,0) \equiv (-4,0,-2P) \equiv (0,2P,0) \bmod \Lambda. \qquad (9.11)$$

No matter what the value of t, we have $(0,t,0) \equiv (0,t',0) \bmod \Lambda$ for some $t' \in [0,2]$. Hence, mod Λ, all points of $\Phi(C)$ lie on the same segment having length 2 and parallel to $(0,1,0)$. $\square$

In our next proof we say that S *contains T mod Λ* if every point of T is Λ-equivalent to a point of S.

Lemma 9.8 *In each case above, we can take $W = C_0$.*

Proof: We consider the vertical P case. The vertical Q case is entirely similar. Let S be the segment with endpoints $W \pm (0,0,1)$ that contains $\Phi(C)$ mod Λ. Let $\zeta = \Phi(C_0)$, the image of the tracker. By definition ζ is equivalent mod Λ to some point on S. Hence S is equivalent mod Λ to a segment S' of length 2 that contains ζ and is parallel to $(0,0,1)$.

Let $\zeta' \in S'$ be some point that is equivalent mod Λ to a point of $\Phi(C)$. If $\|\zeta - \zeta'\| > 1$ we can replace ζ' by $\zeta'' = \zeta' \pm (0,0,2)$ to find an equivalent point on the line through S' that is within 1 of ζ'. Hence, we can arrange that the segment S'' with endpoints $\zeta \pm (0,0,1)$ contains $\Phi(C)$ mod Λ. $\square$

9.5 THE HORIZONTAL CASE

We begin with a result that will simplify our calculations. We say that sets S and T are *translates* mod Λ if there is some vector V such that every point of S is equivalent mod Λ to a point $t + V$ where $t \in T$ and every point of T is equivalent mod Λ to a point $s - V$ for some $s \in S$.

Lemma 9.9 *If C and C' are horizontal trackers of the same type then $\Phi(C)$ and $\Phi(C')$ are translates of each other mod Λ.*

Proof: The proof works the same way independent of the type of the horizontal particle. Let $C = \{c_j\}$ and $C'\{c'_j\}$. Let $\{\widehat{c}_j\}$ and $\{\widehat{c}'_j\}$ be the corresponding points in $\mathbf{R}^3$. The correspondence is as in §9.3.

The key observation is that every horizontal particle has an instance that lies in the western boundary of a block. Cycling the indices we can assume that c_0 and $\widehat{c}_0$ track these instances. We then have one of the following two options:

$$c_j - c'_j = (k\omega, \ell), \qquad c_j - c'_j = (k\omega, \ell \pm \omega). \qquad (9.12)$$

Here $k, \ell \in \mathbf{Z}$ and the sign in the second option depends on the parity of j. Either case leads to

$$\widehat{c}_j - \widehat{c}'_j = (k, \ell, 0). \qquad (9.13)$$

The right-hand side is independent of j. Hence, by Lemma 9.4 the sets $\Phi(C)$ and $\Phi(C')$ are translates mod Λ. $\square$

In the proofs of the next two lemmas we keep track of one extra point because it will help us with the final lemma.

Lemma 9.10 *Let C be a horizontal P tracker. There is some vector W such that mod Λ the image $\Phi(C)$ lies in a segment with endpoints $W \pm (P, P, P)$.*

Proof: We just need to check this in a single case. We will use the (dark) half-particle which has an instance at $(0, 0)$ in the plane.

Let $\{x\}$ denote the half-integer nearest x. We have

$$\{x\} = x - [x] + 1/2, \qquad (9.14)$$

where $[x]$ is the number in $(0, 1)$ representing x mod $\mathbf{Z}$. When $x \in \mathbf{Z}$ we have to make an explicit choice for $\{x\}$ and $[x]$. To make our argument run smoothly, we define

$$\{0/P\} = 1/2, \qquad [0/P] = 0, \qquad \{2p/P\} = \omega - 1/2, \qquad [2p/P] = 1. \qquad (9.15)$$

The instances in our half-particle (and one additional point) are $z_0, ..., z_{2p}$, where

$$z_k = (k/P - k\widehat{\tau}\omega, 0). \qquad (9.16)$$

The extra point z_{2p} corresponds to the partner half-particle of type Q. The center $c_k \in C$ corresponding to z_k is

$$c_k = (\{k/P\} - k\widehat{\tau}\omega, 1/2 \pm \omega). \tag{9.17}$$

The corresponding point in $\boldsymbol{R}^3$ is

$$\widehat{c}_k = (\{k/P\}, 1/2, k). \tag{9.18}$$

Hence

$$\widehat{c}_k - \widehat{c}_0 = (\{k/P\} - 1/2, 0, k). \tag{9.19}$$

Equation 9.5 then gives for $k = 1, ..., 2p$,

$$\Phi(c_k) - \Phi(c_0) = \Phi_0(\widehat{c}_k - \widehat{c}_0) \equiv$$

$$(\mu_k - 2k, \mu_k, \mu_k) \equiv (\nu_k, \nu_k, \nu_k) \mod \Lambda, \tag{9.20}$$

where

$$\mu_k = 2P\{k/P\} - P = 2k - 2P[k/P], \tag{9.21}$$

and

$$\nu_k = \mu_k - 2k = -2P[k/P] \in [-2P, 0]. \tag{9.22}$$

Our result follows immediately. $\square$

Lemma 9.11 *Let C be a horizontal Q tracker. There is some vector W such that mod Λ the image $\Phi(C)$ lies in a segment with endpoints $W \pm (Q, 0, 0)$.*

Proof: The argument is very similar to the previous case. We make the same conventions as in Equation 9.15 with q, Q in place of p, P. We get

$$\widehat{c}_k = (\{k/Q\}, 1/2, 2\omega - k\}, \quad k = 0, ..., 2q. \tag{9.23}$$

Here we are listing the points in reverse order to simplify the equations. Using Equation 9.9, we have

$$\Phi(c_k) - \Phi(c_0) = \Phi_0(\widehat{c}_k - \widehat{c}_0) = \Phi_0(\{k/Q\} - 1/2, 0, -k) \equiv$$

$$(-2Q\{k/Q\} - Q + 2k, 0, 0) \equiv (2Q[k/Q], 0, 0) \mod \Lambda. \tag{9.24}$$

Since $[k/Q] \in [0, 1]$, the conclusion of the lemma follows immediately. $\square$

Lemma 9.12 *In each case above, we can take $W = C_0$.*

Proof: We treat the horizontal P case first. Let S be the segment with endpoints $W \pm (P, P, P)$ that contains $\Phi(C)$ mod Λ. Let $\zeta = \Phi(C_0)$, the image of the tracker. An examination of the case $k = 2p$ of Lemma 9.10 shows that S is the minimal segment with this property. We want to show that ζ must be the midpoint of S. This time the argument is different because $(2P, 2P, 2P) \notin \Lambda$. However, we can use this fact to our advantage.

Our argument refers to the points $\{c_k\}$ from the proof of Lemma 9.10. We have $C_0 = c_p$. By symmetry,

$$\Phi_0(\widehat{c}_{p-k} - \widehat{c}_0) + \Phi_0(\widehat{c}_{p+k} - \widehat{c}_0) = 0. \tag{9.25}$$

This means that

$$(\Phi(c_{p-k}) - \zeta) + (\Phi(c_{p+q}) - \zeta) \in \Lambda. \tag{9.26}$$

The vector on the left-hand side is (t, t, t) for some $t \in [-4P, 4P]$. Since $P < 1$ we have $|t| < 4$. But Λ contains no nonzero vectors having the first coordinate in $(-4, 4)$. This shows that ζ is halfway between $\Phi(c_{p\pm k})$. Since this holds for all $k = 0, ..., p$ and S is minimal, we see that ζ is the midpoint of the segment S.

The type Q case is similar. This time the vector in Equation 9.26 has the form $(t, 0, 0)$ with $|t| \le 4Q < 8$. If this vector lies in Λ then either $t = 0$ or $t = 4$. But $(4, 0, 0) \notin \Lambda$. Hence $t = 0$ and we get the same result as in the type P case. $\square$

Chapter Ten

The Vertical Lemma

10.1 CHAPTER OVERVIEW

We keep the notion from the previous chapter. Referring to Equation 9.1, we have from the previous chapter that $\Phi(\Pi_V) \subset X_V \bmod \Lambda$. In this chapter we study the image $\Phi(\Pi_V)$ more carefully.

We say that $c \in \Pi_V$ has *type* (P, E) if $W\square_c$ contains a light point of type P whose transverse direction points east. Here $\square_c$ is the unit integer square having c as its center and $W\square_c$ is the west edge of $\square_c$. We make the same definitions for the pairs (P, W), (Q, E), and (Q, W). We say that c has *type* $(P, \diamond)$ if $W\square_c$ has a dark point of type P. Note that we cannot assign a direction to such a point. Every $c \in \Pi_V$ has two types, one $(P, *)$ and one $(Q, *)$.

We say that a *container* for (P, E) is a subset $\Sigma \subset X_V$ such that

1. If c has type (P, E) then $\Phi(c) \in \Sigma \bmod \Lambda$.

2. If c has type $(P, \diamond)$ then $\Phi(c) \notin \Sigma \bmod \Lambda$.

We make the same definition for (P, Q), (Q, E), and (Q, W). Here is the main result of the chapter.

Lemma 10.1 (Vertical) *There exists a container $\mathcal{C}$ which works simultaneously for (P, E) and (Q, W). This container is the union of two triangles with vertices*

$$(-2 + P, 1, 1), \qquad (-1 + P, 1, 1), \qquad (-2 + P, 0, 0)$$

and

$$(P, -1 + P, -1 + P), \qquad (-1 + P, -1 + P, -1 + P), \qquad (P, P, P).$$

There is also a container that works simultaneously for (P, W) and (Q, E). This container is obtained by translating $\mathcal{C}$ by $(2, P, P)$.

Figure 10.1 shows the containers $\mathcal{C}$ (lightly shaded) and $\mathcal{C} + (2, P, P)$. We are projecting the picture into the XY plane. We have also drawn a $4 \times 2P$ rectangle around these sets.

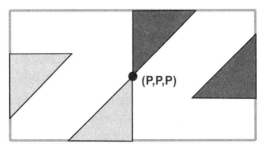

Figure 10.1: The containers $\mathcal{C}$ and $\mathcal{C} + (2, P, P)$.

In §10.2, we will use symmetry to cut down the amount of work we have to do. By symmetry we will only have to look at the case (P, E).

In §10.3, we will translate the picture so that we are trying to prove something more symmetric. The Vertical Lemma associates a point at the center of the unit integer square with a point on the east edge of the square. We get a more symmetric picture when we work with the centers of the east edges rather than the centers of the squares.

In §10.4, we present some alternative formulas for the map $\Phi : \Pi \to X'$ and $\Phi : \Pi \to X$ which will help with our calculations.

In §10.5 we prove the version of the Vertical Lemma for Λ'.

In §10.6, we will make a careful study of the congruences and show that the Λ'-based result implies the Λ-based result.

Computer Tie-In: To see the Vertical Lemma in action, open up the *plaid PET* window and the *plaid PET* control panel. Turn off all the displays on the *plaid PET* control panel to get a blank plotting surface. Next, open up the *sanity checks* control panel. Select the *images by type* option. On the navy blue control panels beneath the *sanity checks* control panel, the settings should be as follows *slice: ver*, *offset: square*, *plot choice: symmetric*. Once you have these settings, you can select a point type and see the corresponding image plotted in the *plaid PET* window. If you want to see a more symmetric picture, change the *offset* option to *edge*. This will reveal the construction in §10.3. Use a fraction like 21/34 to get a nice picture.

10.2 USING SYMMETRY

Lemma 10.2 *The same container works for the types (P, E) and (Q, W) simultaneously. Likewise, the same container works for the types (P, W) and (Q, E) simultaneously. Finally, if the container $\mathcal{C}$ works for the type (P, E) then the translation $\mathcal{C} + (2, P, P)$ works for (P, W).*

Proof: Suppose $c \in \Pi$ lies on a horizontal midline H. Reflection in H interchanges the two light points on $W\square_c$, reverses their types, and reverses their directions. Hence the two types (P, E) and (Q, W) always appear together in this situation, and likewise for the types (P, W) and (Q, E).

The map $(x, y) \to (x, y) + (0, \omega)$ preserves the types of the particles and reverses their directions. Also,

$$\Phi(x, y_+(0, \omega)) = \Phi(x, y) + (2, P - 2, P - 2) \equiv \Phi(x, y) + (2, P, P) \bmod \Lambda.$$

This completes our proof. □

In view of these symmetry results, it suffices to prove the Vertical Lemma for the case (P, E).

10.3 TRANSLATING THE PICTURE

We modify the original construction. Let Π' denote the set of centers of unit vertical segments in $\mathbf{R}^2$. We obtain Π' from Π by subtracting $(1/2, 0)$ from every point.

We define the *type* of $c' \in \Pi'$ to be (P, E) if there is a light point on the unit vertical segment containing c' which has type P and transverse direction pointing east. By definition c' and $c = c' + (1/2, 0)$ have the same types. The advantage of working with the new definition is that c' is more directly related to the light points on it.

We say that a *modified container* for (P, E) is a subset $\Sigma \subset X_V$ such that

1. If $c' \in \Pi'$ has type (P, E) then $\Phi(c) \in \Sigma \bmod \Lambda$.

2. If $c' \in \Pi'$ has type $(P, \diamond)$ then $\Phi(c) \notin \Sigma \bmod \Lambda$.

Here is a reformulation of the Vertical Lemma for the type (P, E).

Lemma 10.3 *There exists a container for (P, E) which is a union of the 2 special triangles having vertices*

$$(-2, 1 - P, 1 - P), \qquad (-1, 1 - P, 1 - P), \qquad (-2, -P, -P)$$

and

$$(0, -1, -1), \qquad (-1, -1, -1), \qquad (0, 0, 0).$$

The equivalence of the two lemmas comes from the fact that when $c = c' + 1/2$,

$$\Phi(c') - \Phi(c) = \Phi(-1/2, 0) = (-P, -P, -P).$$

The picture of the modified container is exactly like the lightly shaded part of Figure 10.1, except that the point labeled (P, P, P) is now labeled $(0, 0, 0)$.

10.4 SOME USEFUL FORMULAS

Let $[x]_{2n}$ denote the representative of $x \bmod 2n\mathbf{Z}$ in $[-n, n)$.

Lemma 10.4 $\Phi(x, y) = (T, U_1, U_2) \bmod \Lambda$, *where*

- $T = [2Px + 2y]_4$.

- $b = \frac{1}{2}PT(x, y)$.

- $U_1 = [PQx + b - Py]_2$.

- $U_2 = [PQx + b + Py]_2$.

Proof: Let Φ^* be the map given by (T, U_1, U_2). We want to show that $\Phi^* = \Phi \mod \Lambda$. To save words, we think of the ranges of these maps as $\mathbf{R}^3/\Lambda$, so we are just trying to prove that $\Phi = \Phi^*$.

The first coordinate is easy: $T(c)$ is just the expression for the first coordinate of $\Phi(c)$. Hence, $\Phi = \Phi^*$ in the first coordinate. Near the origin, we have

$$U_1(x, y) = PQx + P^2x + Py - Py = Px(P + Q) = 2Px = \Phi_2(x, y).$$

$$U_2(x, y) = PQx + P^2x + Py + Py = Px(P + Q) = 2Px + 2Py = \Phi_3(x, y).$$

Here Φ_2 and Φ_3 are the second and third coordinates of Φ. This calculation shows that $\Phi = \Phi^*$ in a neighborhood of the origin and that Φ and Φ^* have the same linear part.

We know that Φ is a locally affine map on all of $\mathbf{R}^2$. We will show that Φ^* is also a locally affine map on all of $\mathbf{R}^2$. Since $\Phi = \Phi^*$ on an open set and both maps are locally affine, we must have $\Phi = \Phi^*$. So, it only remains to check that Φ^* is locally affine. Since the linear part of Φ^* is constant, it suffices to prove that Φ^* is locally affine away from a countable set of lines and everywhere continuous.

Consider Φ^* near a point $(x, y) \in \mathbf{R}^2$ that does not lie in the family of lines given by $2Px + 2y \in 4\mathbf{Z} + 2$. Both $T(x, y)$ and $b(x, y)$ vary linearly near (x, y). The coordinates $U_j(x, y)$ could jump by 2 units, but they vary linearly mod 2. Since Λ contains the vectors $(0, 2, 0)$ and $(0, 0, 2)$ we see that Φ^* is locally affine near (x, y).

To check global continuity of Φ^* we just have to study what happens when points (x, y) and (x', y') are close together and on either side of a line of the form $2Px_2y \in 2\mathbf{Z} + 2$. In this case (after suitably ordering the points) we have $T \approx 2$ and $T' \approx -2$. This gives $b \approx P$ and $b' \approx -P$. But then

$$\Phi^*(x, y) - \Phi^*(x', y') \approx (4, 2P, 2P) \in \Lambda.$$

This shows that Φ^* is globally continuous. $\square$

Remark: If y is a half-integer, and we work mod Λ', then we get the same formula, except that we can use the simpler expression

$$T(x, y) = [2Px + 1]_2. \tag{10.1}$$

in place of $[2Px + 2y]_4$

Let $\mathbf{Z}_0$ and $\mathbf{Z}_1$ denote the sets of even and odd integers respectively. We define the following 4 functions:

- $F_H(x,y) = [2Py]_2$

- $F_V(x,y) = [2Px]_2$

- $F_P(x,y) = [Py + P^2x + 1]_2$

- $F_Q(x,y) = [Py + PQx + 1]_2$

We shall never consider expressions of the form $[k]_2$ when $k \in \mathbf{Z}_1$, so there is never any ambiguity in the definition.

Lemma 10.5 *Let R stand for either P or Q. A vertical intersection point z lying on a slanting line of slope $-R$ is a light point if and only if $F_R(z)$ and $F_V(z)$ have the same sign and $|F_R(z)| < F_V(z)$.*

Proof: The value $\omega F_V(z)$ computes the signed capacity of the vertical line containing z. The function F_R is constant on lines of slope $-R$ and $\omega F_R(z)$ computes the signed mass of the slanting line of slope $-R$ through V. $\square$

10.5 THE UNDIRECTED RESULT

Figure 10.2 shows the image of the container from Lemma 10.3 under the quotient map $X \to X'$. Again, we are projecting to the XY plane.

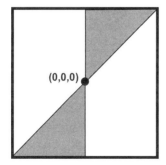

Figure 10.2: The container in the quotient space.

This is the set of points $(T, U, U) \in [-1, 1]^3$ such that $UT \geq 0$ and $|T| \leq U$. Notice that this condition is fairly close to our first definition of a light point given in §1.5. We will prove that the union of two triangles in Figure 10.2 is a modified container for light points of type P when we work mod Λ'. That is, if c is the center of a vertical unit integer segment then $\Phi(c)$ lies in this set mod Λ if and only if one of the light points on this edge has type P.

Let $c \in \Pi_V$. We have $c = (x, \pm\omega/2)$. For ease of notation we consider only the case when $c = (x, \omega/2)$. The other case has an entirely similar treatment. Let $z = (x, y)$ denote the intersection of a slanting line of type P with the vertical unit integer segment centered at c. We use the notation from the previous section.

Lemma 10.6 $y = \omega/2 - T(c)/2$.

Proof: We have
$$y \in (\omega/2 - 1/2, \omega/2 + 1/2). \tag{10.2}$$
The point z lies on a line of slope $-P$ which has an integer y-intercept. Hence $Px + y \in \mathbf{Z}$. By construction, $Px + y$ is the integer nearest $Px + \omega/2$. Hence
$$Px + y = Px + \omega/2 - [Px + \omega/2]_1.$$
Subtracting off Px, we get
$$y = \omega/2 - [Px + \omega/2]_1. \tag{10.3}$$
Since $[t]_1 = [2t]_2/2$, we get
$$y = \omega/2 - \frac{1}{2}[2Px + \omega]_2 =^* \omega/2 - \frac{1}{2}[2Px + 1]_2 = \omega/2 - T(c)/2. \tag{10.4}$$
The starred equality uses the fact that $\omega - 1 \in 2\mathbf{Z}$. $\square$

Now we are going to do a trick. Rather than work with the point z we will work with the point
$$z^* = (x, \omega/2 + T(c)/2). \tag{10.5}$$
Reflection in the horizontal midline of the block swaps z and z^*. By symmetry, z is a light point of type P if and only if z^* is a light point of type Q.

From the remark in the previous section, we have
$$F_V(z^*) = [2Px]_2 = [T(c) + 1]_2. \tag{10.6}$$

Lemma 10.7
$$F_Q(z^*) = [U_1(c) + 1]_2. \tag{10.7}$$

Proof: We have
$$z^* = (x, y^*), \quad y^* = \omega/2 + T(c)/2, \qquad c = (x, \omega/2).$$
Using Lemma 10.4 mod Λ',
$$F_Q(z^*) = [Py^* + PQx + 1]_2 =$$
$$[P\omega/2 + PT(c)/2 + PQx + 1]_2 =$$
$$[PQx + b(c) + P\omega/2 + 1]_2 =$$
$$[U_2(c) + 1]_2 = [U_1(c) + 1]_2. \tag{10.8}$$
The last equality comes from the fact that we already know $U_1(c) = U_2(c)$. $\square$

From Equations 10.6 and 10.7 we see that

- $F_V(z^*)$ and $F_Q(z^*)$ have the same sign iff $T(c)$ and $U_1(c)$ have the same sign.

- $|F_Q(z^*)| < |F_V(z^*)|$ iff $|T(c)| < |U(c)|$.

Hence z^* is a light point of type Q iff z is a light point of type P iff $\Phi(c)$ lies in the container from Figure 10.2. This completes the proof of Lemma 10.3 in the (undirected) Λ' case.

10.6 DETERMINING THE DIRECTIONS

Now we deduce the directed result from the undirected result. We keep the same notation as in the previous section but work mod Λ. The first coordinate of $\Phi(c)$ is

$$[2Px + \omega]_4. \tag{10.9}$$

To deduce the directed result from the undirected result, it suffices to show that this quantity is positive when z has type (P, E) and negative when z has type (P, W).

The vertical line containing c and z either has positive sign or negative sign. We first consider the case when this line has positive sign. This is to say that $[4px]_{2\omega} > 0$. Dividing through by ω, we get

$$[2Px]_2 > 0. \tag{10.10}$$

Let y_0 denote the y-intercept of the line of slope $-P$ through z. By the argument in Lemma 1.3, the point z has type (P, E) iff y_0 is even. We have

$$y_0 = Px + y.$$

y_0 is the integer nearest $Px + \omega/2$. So, the integer nearest $Px + \omega/2$ is even. Hence the even integer nearest $2Px + \omega$ is congruent to 0 mod 4. In summary,

1. $[2Px]_2 > 0$.

2. The even integer nearest $2Px + \omega$ is congruent to 0 mod 4.

Figure 10.3: How the constraints force the sign.

The left half of Figure 10.3 explains why these two conditions imply that $[2Px+\omega]_4 > 0$. The thick lines above the interval show the constraints given by our first condition and the thick lines below show the constraints given by the second.

When our vertical line has negative sign, the argument is similar. The right side of Figure 10.3 shows this case.

Chapter Eleven

The Horizontal Lemma

11.1 CHAPTER OVERVIEW

In this chapter we do for Π_H what we did for Π_V in the last chapter. We make all the same definitions. We use the south edges instead of the west edges. So, for instance, $c \in \Pi_H$ has type (Q, N) if $S\square_c$ contains a light point of type Q whose transverse direction points north. Note that c again has two types, one of type $(P, *)$ and one of type $(Q, *)$. Indeed, the intersection point in $S\square_c$ associated to $c \in \Pi_H$ is always a double point lying at the center of a block.

Lemma 11.1 (Horizontal) *There exists a container $\mathcal{C}$ which works simultaneously for (P, N) and (Q, S). This container is the union of two triangles with vertices*

$$(1, 1, -1 + P), \qquad (1, 0, -1 + P), \qquad (1, 1, P)$$

and

$$(-1, 1 - P, 1), \qquad (-1, 2 - P, 1), \qquad (-1, 1 - P, 0).$$

There is also a container that works simultaneously for (P, S) and (Q, N). This container is obtained by translating $\mathcal{C}$ by $(2, P, P)$.

Figure 11.1 shows $\mathcal{C}$ projected into the YZ plane.

Figure 11.1: The set $\mathcal{C}$.

Our proof follows a similar outline as in the previous chapter. In §11.2 we will use symmetry to reduce the Horizontal Lemma to a simpler statement. In §11.3, we modify the construction as in the vertical case, reducing the Horizontal Lemma to the simpler Lemma 11.4. In §11.4 we prove two easy technical lemmas. In §11.5 we prove the Λ' version of Lemma 11.4. In §11.6 we keep track of the signs and prove Lemma 11.4 as stated.

Computer Tie-In: Same as the previous chapter, except set *slice: hor*.

11.2 USING SYMMETRY

Lemma 11.2 *The same container works for the types (P, S) and (Q, N) simultaneously. Likewise, the same container works for the types (P, S) and (Q, N) simultaneously.*

Proof: As discussed in §1.5 (the Midline Case), the edge $S\square_c$ contains two intersection points at its midpoint z when c is contained in a vertical midline of a block. One could say that both these points have both types P and Q. What is important here are the directions.

Let L_P and L_Q respectively denote the slanting lines of slope $-P$ and $-Q$ through z. These lines have the same sign because z is a light point with respect to both. Let L'_Q be the line of slope Q through z. In Lemma 1.3 we proved that L'_Q and L_Q give the same direction to z. Also, from the second definition of light points given in §1.5, the lines L_Q and L'_Q have the same sign. Hence L_P and L'_Q have the same sign. Since the first coordinate of z has the form $k\omega + \omega/2$ and since $P + Q = 2$, we see that the difference in the Y-intercepts of L_P and L'_Q is an odd integer. Hence the Y-intercepts of L_P and L'_Q have opposite parity. As the proof of Lemma 1.3 shows, these lines give the opposite transverse direction to z. So, either c has types (P, N) and (Q, S) or c has types (P, S) and (Q, N). $\square$

In the horizontal case, we will use symmetry in a slightly different way. Referring to the Horizontal Lemma, let $\sigma_\pm$ and $\nu_\pm$ respectively denote the triangles of $\mathcal{C}$ and $\mathcal{C} + (2, P, P)$ that are contained in the slice $(\pm 1, *, *)$. Let $\Pi_H^\pm$ denote the subset of Π_H consisting of points c such that $\Phi(c) = (\pm 1, *, *)$ mod Λ. We can redefine the notion of container with respect to $\Pi_H^\pm$. To distinguish what we have already done, we use the words *positive container* and *negative container*. Thus, for example, σ_+ is a *positive container* for (P, S) if the following holds:

1. If $c \in \Pi_H^+$ has type (P, S) then $\Phi(c) \in \sigma_+$ mod Λ.

2. If $c \in \Pi_H^+$ has type $(P, \diamond)$ then $\Phi(c) \notin \Sigma$ mod Λ.

The following lemma allows us to reduce the problem to Π_H^*.

Lemma 11.3 *If σ_+ and ν_+ are respectively positive containers for the types (P, S) and (P, N) then they are respectively negative containers for (P, S) and (P, N) as well.*

Proof: One can check from the definitions of our sets in the Horizontal Lemma that

$$\sigma_- = \nu_+ \pm (2, P, P), \qquad \nu_- = \sigma_+ \pm (2, P, P) \qquad \text{mod } \Lambda. \tag{11.1}$$

Either sign in front of $(2, P, P)$ works because $(4, 2P, 2P) \in \Lambda$.

Let $c \in \Pi_H^-$. Let $z \in S\square_c$ be the corresponding point of type $(P, *)$. Let
$$z' = z + (0, \omega), \qquad c' = c + (0, \omega).$$
Then $z' \in S\square_{c'}$ and
$$\Phi(c) = \Phi(c') \pm (2, P, P) \bmod \Lambda.$$
Hence $c' \in \Pi_H^+$. Finally, we note that translation by $(0, \omega)$ preserves the (undirected) types and reverses the directions.

It only remains to put everything together.

- If c has type $(P, \diamond)$ then c' has type $(P, \diamond)$. By hypotheses, $\Phi(c') \notin \sigma_+$ mod Λ. Hence $\Phi(c) \notin \nu_-$ mod Λ. Likewise $\Phi(c') \notin \nu_+$ mod Λ. Hence $\Phi(c) \notin \sigma_-$ mod Λ.

- If c has type (P, S) then c' has type (P, N) and $\Phi(c') \in \nu_+$ mod Λ. Hence $\Phi(c) \in \sigma_-$ mod Λ.

- If c has type (P, N) then c' has type (P, S) and $\Phi(c') \in \sigma_+$ mod Λ. Hence $\Phi(c) \in \nu_-$ mod Λ.

This completes the proof. $\square$

11.3 TRANSLATING THE PICTURE

We modify the construction made in §10.3 using horizontal instead of vertical unit integer segments. This time let Π' denote the set of centers of unit horizontal segments. We associate $c' \in \Pi'$ to a horizontal intersection point z iff $c \in \Pi$ is associated to z and $c' = c - (0, 1/2) \in S\square_c$. So, c' and z lie in the same unit horizontal edge. In this way we assign types to points in Π', and in particular to points in Π_H'. In fact, when $c \in \Pi_H'$, we have $c' = z$ because z lies in the vertical midline of a block.

With the definitions in place, we define *modified containers* just as in §10.3. Note that
$$\Phi(c') \equiv \Phi(c) - (1, 0, P) \bmod \Lambda. \qquad (11.2)$$
To formulate our modified result, we set
$$\sigma_+' = \sigma_+ - (1, 0, P), \qquad \nu_+' = \nu_+ - (1, 0, P) - (0, 2, 0). \qquad (11.3)$$
(There is no harm in subtracting off $(0, 2, 0)$ because this vector belongs to Λ; we do it for cosmetic purposes.)

σ_+' is the triangle with vertices
$$(0, 1, -1), \qquad (0, 0, -1), \qquad (0, 1, 0). \qquad (11.4)$$
We get μ_+' by adding $(2, P, P)$ to the second triangle in the Horizontal Lemma, then subtracting $(1, 2, P)$. The vertices are
$$(0, -1, 1), \qquad (0, 0, 1), \qquad (0, -1, 0). \qquad (11.5)$$
Figure 11.1 shows these two triangles. The lightly shaded triangle is σ_+' and the darkly shaded triangle is μ_+'. The square frame is $[-1, 1]^2$. We are drawing the picture in the slice $(0, *, *)$.

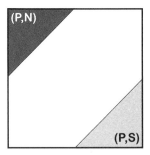

Figure 11.2: σ'_+ and μ'_+.

We have reduced the Horizontal Lemma to the following result.

Lemma 11.4 *Suppose that $c' \in \Pi'_H$ and that $\Phi(c') \in (0, *, *)$ mod Λ. Then*

- *If c' has type (P, S) then mod Λ the point $\Phi(c')$ lies in the triangle with vertices $(0, 1, -1)$, $(0, 0, -1)$, $(0, 1, 0)$..*

- *If c' has type (P, N) then mod Λ the point $\Phi(c')$ lies in the triangle with vertices $(0, -1, 1)$, $(0, 0, 1)$, $(0, -1, 0)$..*

- *If c' has type $(P, \diamond)$ then mod Λ the point $\Phi(c')$ lies in neither of the triangles just mentioned.*

11.4 TWO EASY TECHNICAL LEMMAS

We use the notation from §11.4. In particular, $[x]_{2n}$ denote the representative of x mod $2n\mathbf{Z}$ in $[-n, n)$. We call a pair of points $(a, b) \in \mathbf{R}^2$ a *good pair* if $[a]_2 \leq [b]_2$ and $[a_2]$ has the same sign as $[b_2]$. The set of good pairs in $[-1, 1]$ is exactly the set shown in Figure 10.2.

Lemma 11.5 *Let $b_1 = a_1 - a_2$ and $b_2 = a_1 - 1$. Then (a_1, a_2) is a good pair if and only if (b_1, b_2) is a good pair.*

Proof: Since the affine transformation $T(x, y) = (x - y, x - 1)$ preserves $\mathbf{Z}^2$, the pair (a_1, a_2) satisfies the non-integrality condition if and only if the pair (b_1, b_2) does. If this lemma is true for the inputs (a_1, a_2) it is also true for the inputs $(a_1 + 2k_1, a_2 + 2k_2)$ for any integers k_1, k_2. For this reason, it suffices to consider the case when $a_1, a_2 \in (-1, 1)$. From here, an easy case-by-case analysis finishes the proof. For instance, if $0 < a_1 < a_2$ then $0 > b_1 > b_2$. The other cases are similar. $\square$

Lemma 11.6 *Let R stand for either P or Q. A horizontal intersection point z lying on a slanting line of type R is a light point if and only if $F_P(z)$ and $F_H(z)$ have the same sign and $|F_R(z)| < F_H(z)$.*

Proof: This has essentially the same proof as Lemma 11.6. $\square$

11.5 THE UNDIRECTED RESULT

We work mod Λ'. The point $c' \in \Pi'_H$ is already an intersection point. It lies on slanting lines of slope $\pm P$ and slope $\pm Q$. It is either a double light point or a double dark point. For this reason, the only information we have about c' is that it is either light or dark. Here is the Λ' version of Lemma 11.4.

Lemma 11.7 *Let Υ be the union of two triangles mentioned in Lemma 11.4. Suppose that $c' = (k\omega/2, y)$ for some odd integer k. Then c' is light if and only if $\Phi(c') \in \Upsilon \bmod \Lambda'$.*

Proof: By Lemma 10.4 the coordinates of $\Phi(c') \bmod \Lambda'$ are
$$0, \qquad U_1(c') = [PQx - Py]_2, \qquad U_2(c') = [PQx + Py]_2. \qquad (11.6)$$
Referring to the formulas in §10.4, we have
$$F_P(c') = F_Q(c') = [PQx + Py + 1]_2, \qquad F_H(c') = [2Py]_2. \qquad (11.7)$$
Setting $U_1 = U_1(c')$, etc., we see that mod $2\mathbf{Z}$ we have
$$U_1 = F_P - F_H + 1, \qquad U_2 = F_P + 1. \qquad (11.8)$$
Let $U'_j = U_j + 1 \bmod 2\mathbf{Z}$. By Lemma 11.6, the point c' is a light point if and only if $|F_P| < |F_H|$ and these two quantities have the same sign. By Lemma 11.5, the set Υ' of points (U'_1, U'_2) corresponding to the light points is the same set: It is the set where $|U'_1| < |U'_2|$ and the two quantities have the same sign. But $\Upsilon' = \Upsilon \pm (1,1) \bmod 2\mathbf{Z}$. So, c' is light iff $\Phi(c') + (0,1,1) \in \Upsilon'$ mod Λ' iff $\Phi(c') \in \Upsilon \bmod \Lambda'$. This completes the proof. $\square$

11.6 DETERMINING THE DIRECTIONS

Proof of Lemma 11.4: Suppose that $c' = (x, y) \in \Pi'_H$. By hypotheses, $\Phi(c') = (0, U_1, U_2) \bmod \Lambda$. In particular,
$$[2Px + 2y]_4 = 0. \qquad (11.9)$$
In view of Lemma 11.7, it suffices to prove the following statements under the assumption that c' does not have type $(P, \diamond)$:

1. If $[U_2 - U_1]_2 > 0$ then c' has type (P, N).

2. If $[U_2 - U_1]_2 < 0$ then c' has type (P, S).

We prove the first statement. The second one has the same proof.

By Lemma 10.4, we have
$$[U_2 - U_1]_2 = [2Py]_2 > 0. \qquad (11.10)$$
The sign of the horizontal line L through c' is given by the sign of $[2Py]_2$. Hence c' has positive sign. The proof of Lemma 1.3 shows that L is directed upward iff the Y-intercept $y_0 = y + Px$ is even. Equation 11.9 says that $2y_0$ is divisible by 4. Hence y_0 is even. But then the transverse direction of c' points north. Hence c' has type (P, N) as claimed. $\square$

Chapter Twelve

Proof of the Main Result

12.1 CHAPTER OVERVIEW

In this chapter we will put together the ingredients from the last 3 chapters – the Segment Lemma, the Horizontal Lemma, and the Vertical Lemma – and prove Theorem 8.2. The three technical lemmas do not mention the partition of the space X at all, but they do give a lot of control over how the nature of the particles tracked by points in the plaid grid Π influences the image of such grid points under the classifying map Φ. It remains to compare the three results above to the partition and see that everything matches.

Our proof has a kind of product structure. The Vertical Lemma and the Horizontal Lemma give us good information about the image $\Phi(c')$ when $c' \in \Pi_V \cup \Pi_H$. Given a general $c \in \Pi$ of interest to us, we know that a particle tracker C containing c also has an anchor point $c' \in \Pi_H \cup \Pi_V$. We use the Segment Lemma to glean information about the image of $\Phi(c)$ from the image of $\Phi(c')$.

In the vertical case, we (hereby) extend the notion of a container so that it works for all points of Π and not just those of Π_V. The definition is precisely the same. Likewise, in the horizontal case we hereby extend the notion of a container so that it works for all points of Π and not just those of Π_H. In §12.2 we carry out the program discussed in the previous paragraph to show that these containers are each a union of two prisms.

In §12.3 we discuss some extra symmetry of our partition which will make our arguments go more smoothly. In §12.5 we compare the prism containers in the vertical case to the partition of X and deduce that Theorem 8.2 is true for the vertical unit integer segments. In §12.4 we compare the prism containers in the vertical case to the partition of X and deduce that Theorem 8.2 is true for the horizontal unit integer segments. The two results together finish the proof.

12.2 PRISM STRUCTURE

For us, a *prism* is the affine image of a product $\Delta \times I$, where Δ is a triangle and I is an interval.

In the next result, the prisms are such that Φ never maps a point into their boundaries. (This follows from Lemma 8.1 and from the fact that the prism boundaries lie in the boundaries of our partition.) So, the fact that their boundaries can intersect is irrelevant.

Lemma 12.1 (Prism) *For every parameter and every type there is a container which is a union of 2 prisms. The vertices of this prism vary linearly with the parameter. The Λ orbit of these prisms consists of an infinite family of prisms whose interiors are pairwise disjoint.*

Proof: We will give the proof in detail for the type (P, N) and then discuss why the same argument works for all types. Before we start, we note that $Q = 2 - P$. We will use $2 - P$ in place of Q for cosmetic purposes.

Let Σ denote the the union

$$\Sigma = \bigcup_{\zeta \in \Sigma_0} I_\zeta, \tag{12.1}$$

where I_ζ is the segment with endpoints $\zeta \pm (2 - P, 0, 0)$. By the Segment Lemma, $\Phi(C) \subset \Sigma$ mod Λ. Note that the segments are transverse to Σ and inject in X mod Λ. Since the interiors of Σ_0 are disjoint in X, the interiors of the two components of Σ_2 are disjoint in X. So, when we lift to $\mathbf{R}^3$ we get an infinite Λ-invariant family of prisms which have pairwise disjoint interiors.

Now we show that Σ is a container for the type (P, N). Let C be a tracker for a horizontal half-particle whose type is P. Suppose first that C tracks a light particle whose transverse direction is north. Let C_0 be the anchor of C. Let Σ_0 be the container associated to (P, N) from the Horizontal Lemma. Σ_0 is a union of two triangles, one in the plane $(1, *, *)$ and one in the plane $(-1, *, *)$. If C tracks a dark particle, then $\Phi(C_0) \notin \Sigma_0$ mod Λ. But then the Segment Lemma says that no point of $\Phi(C)$ lies in Σ mod Λ. Hence Σ is a container for the type (P, N).

For the record, one of the prisms has vertices

$$(1, 1, -1 + P) \pm \eta, \quad (1, 0, -1 + P) \pm \eta, \quad (1, 1, P) \pm \eta, \quad \eta = (2 - P, 0, 0). \tag{12.2}$$

The other cases have similar treatments. The important points are that the segment I_ζ is always transverse to Σ_0, that the interior of each segment I_ζ injects into X when we quotient out by Λ, and that the interior of Σ_0 injects onto 2 disjoint open triangles when we quotient out by Λ. These properties guarantee that the lift of Σ to $\mathbf{R}^3$ in all cases is an infinite family of prisms with pairwise disjoint interiors. This completes the proof. □

So far, we have been taking slices one parameter at a time. For each type, we produce a container Σ_P for each parameter $P = 2A/(1 + A)$. Now we take the grand union

$$\Sigma = \bigcup_{P \in [0,1]} \Sigma_P \times \{P\} \subset \mathbf{R}^4. \tag{12.3}$$

Given the way that the vertices of Σ_P vary with P, the set Σ is a convex integral polytope! The key phenomenon behind this is that the vertices of the prisms vary linearly and the normals to the faces do not change with the parameter.

We find the vertices of the polytope by setting $P = 0$ and $P = 1$ in the formulas. There are no vertices in the intermediate slices because the combinatorics of the prism does not change.

In the example from Equation 12.2, the 12 vertices are

$$(1, 1, -1, 0) \pm (2, 0, 0, 0), \quad (1, 0, -1, 0) \pm (2, 0, 0, 0), \quad (1, 1, 0, 0) \pm (2, 0, 0, 0),$$

$$(1, 1, 0, 1) \pm (1, 0, 0, 0), \quad (1, 0, 0, 1) \pm (1, 0, 0, 0), \quad (1, 1, 1, 1) \pm (1, 0, 0, 0).$$

The grand union of all the triangles in Equation 12.2 is the convex hull of these 12 vertices.

The affine PET on X gives us a partition of X^2 into 4-dimensional convex integral polytopes. In this section we have constructed a union of 16 convex integral polytopes which determine the nature of the intersection points. The rest of the proof of Theorem 8.2 amounts to lining up these two collections and seeing that they specify the same information.

Notice that the collection of containers produced here is not a partition of X. For instance, there are some point $c \in \Pi$ whose unit integer square consists entirely of dark intersection points. The point $\Phi(c)$ does not lie in any of the containers. Indeed, c has the "all-dark" property if and only if $\Phi(c)$ does not lie in any other containers.

Notice also that the containers can overlap. For instance, there are plenty of grid points c such that $W\square_c$ contains a light point of type (P, W) and a light point of type (Q, E). The corresponding containers overlap. However, their intersection is contained in the union of partition polytopes labeled by $\emptyset$. This is one example of how the two collections will line up perfectly to prove Theorem 8.2.

12.3 SOME EXTRA SYMMETRY

We are going to give arguments below that Theorem 8.2 is true with respect to the west and south edges of the unit integer squares. Thus, for instance, if $W\square$ is the west edge of some unit integer square, there is one light point on $W\square$ if and only if the master tiling in $\square$ involves W. In this section we explain how to deduce Theorem 8.2 for the east and north edges once we know it for the west and south edges. We exploit the symmetry of the situation.

First of all, everything in sight is invariant under the translation

$$\tau(x, y) = (x + \omega^2, y).$$

This map carries intersection points to intersection points and preserves the shades and transverse directions. Likewise $\Phi \circ \tau = \Phi$, where Φ is the classifying map. We extend both our constructions, the plaid construction and the master tiling, to all of $\boldsymbol{R} \times [-\omega, \omega]$. Once we know Theorem 8.2 for the west and south edges in the fundamental blocks, we know it for the west and south edges in the entire infinite strip.

Let ρ_2 and ρ_3 denote reflection in the origin in $\boldsymbol{R}^2$ and $\boldsymbol{R}^3$ respectively. The map ρ_2 is a symmetry of the plaid model construction: ρ_2 carries light points to light points and maps transverse directions to transverse directions. Moreover, ρ_2 swaps east edges with west edges and north edges with south edges. What we mean is that ρ_2 maps the east edge of $\square$ to the west edge of $\rho(\square)$. Likewise for the other edges.

As we mentioned in §8.3, the map ρ_3 (negation) permutes the polytopes and changes the labels as follows: east and west are swapped and north and south are swapped.

Finally, we observe that $\Phi \circ \rho_2 = \rho_3 \circ \Phi$. From this symmetry we see that ρ_2 is a symmetry of the master tiling. In short, ρ_2 is a symmetry of both constructions, swaps east edges with west edges, and swaps north edges with south edges. Given this symmetry, Theorem 8.2 for the east and north edges follows from Theorem 8.2 for the west and south edges.

12.4 THE VERTICAL CASE

We will use the description of the plaid PET from §8.3. The 4-dimensional version of X is fibered over the (x, P) plane. The (x, P) base space is partitioned into right-angled isosceles triangles which have the property that the fibers over and two points in the same open triangle are combinatorially identical. Call this the *stability property*. Moreover, all such fibers are partitioned into 4×4 grids of rectangles.

From the discussion in the previous section, we just have to worry about the west edges of unit integer squares. Let $W_{\leftarrow}$ denote the union of polytopes which have labels of the form $*W$. These polytopes correspond to tiles in the master tiling which point into the west edge. Likewise, we define $W_{\rightarrow}$ to be the union of polytopes which have labels of the form $W*$. These polytopes correspond to tiles in the master tiling which point out of the west edge. The left-hand side of Figure 12.1 shows these sets over the fiber $(0, 2/5)$.

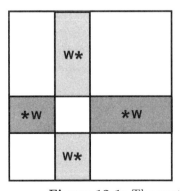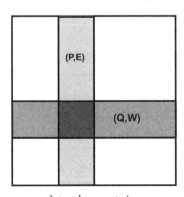

Figure 12.1: The partition compared to the container.

Now let's consider the containers. It follows from Statements 1 and 2 of the Segment Lemma that each vertical container intersects each fiber in a rectangle having one of two types.

1. For type P the rectangle has the form $[-1, 1] \times [a, b]$.

2. For type Q the rectangle has the form $[a, b] \times [-1, 1]$.

By inspection we see that the combinatorics of the intersection does not change over the interiors of the triangles in the base partition. We have the same stability. The right-hand side of Figure 12.1 shows the intersection of the $(0, 2/5)$ fiber with the containers. The pattern is the same, except that the intersection of the two strips is precisely the square omitted on the left-hand side. (The (P, W) and (Q, E) containers do not intersect this fiber.)

Comparing the two pictures we see that Theorem 8.2 holds in the vertical case for this particular parameter and this particular fiber. Here is a breakdown of what can happen in this fiber. If $c \in \Pi$ is a point such that $\Phi(c)$ lands in this fiber, then one of 4 things can happen:

- $\Phi(c)$ lands in the white part of the right-hand side of Figure 12.1. In this case there are no light points on $W \square_c$. At the same time the master tile in $\square_c$ does not involve $W \square_c$.

- $\Phi(c)$ lands in the central dark square on the right side of Figure 12.1. In this case there are 2 light points on $W \square_c$. At the same time the master tile in $\square_c$ does not involve $W \square_c$.

- $\Phi(c)$ lands in the container for (P, E) but not in the container for (Q, W). In this case, $W \square_c$ has one light point and it is directed east. At the same time, $\Phi(c) \in W_\rightarrow$, so the master tile in $\square_c$ points out of c.

- $\Phi(c)$ lands in the container for (Q, W) but not in the container for (P, E). In this case, $W \square_c$ has one light point and it is directed west. At the same time, $\Phi(c) \in W_\leftarrow$, so the master tile in $\square_c$ points into c.

In all cases, the two systems for assigning connectors agree on $W \square_c$.

By direct inspection, we check that such a picture exists in every fiber. This looks like a massive check, but in all cases we are working with slices of convex integer polytopes, and we have the stability property. So, we just have to check the agreement in 3 fibers over each triangle in the base partition. This comes to an inspection of 18 pictures like Figure 12.1. We make this check using our program. Below we will explain an alternate approach which does not require a check like this.

Computer Tie-In: You can make this check for yourself. On the main program open the *plaid PET* window. On the *plaid PET slicer* make the settings as follows.

- *slice: X*,

- *parameter: locked*,

- *PET Choice: neutral*,

- *tiling extent: leftmost*.

After selecting a parameter A you should click the *grab parameter* button to set the slice at $P = 2A/(1 + A)$. I recommend selecting A to have a fairly large denominator, so that you get a fine plot.

Open up the *PET color* window. To match the case considered above, click the big rectangular button labeled W that is slightly more than halfway down on the left side of the *PET color* window. This highlights the union of tiles labeled $*W$ and $W*$.

Next, open up the *sanity checks* control panel and select *images by type*. Select the P *east* and Q *west* particles on the *point type* window. The auxiliary settings should be

- *slice: ver*,

- *offset: square*,

- *plot choice: all*.

Click around the *Plaid PET slicer* to see plots of the kind shown in Figure 12.1 for various slices.

Alternate Proof: There is another way to think about the proof which removes the need for us to make a visual inspection of the polytopes. Instead we can look at the two tilings. In order to verify the equalities in general, it suffices to verify them for 2 parameters which have the property that a point is mapped into a fiber above each open triangle in the partition. The parameters $2/5$ and $3/8$ would do the job. If there was some mismatch between the two pictures then by Lemma 8.1 there would be some $c \in \Pi$ where there was a mismatch in some $W\square_c$ between the two schemes. We check directly that there are no such mismatches for these parameters.

12.5 THE HORIZONTAL CASE

From the discussion in §12.3 it suffices to consider the south edges of unit integer squares. We analyze things in the horizontal case using the same fibration picture as in the vertical case. Figure 12.2 shows the same fiber as in the vertical case, namely the fiber over $(0, 2/5)$. The black region on the right-hand side is where the various containers overlap. Inspecting this picture we see once again that both schemes agree on the edge $S\square_c$ provided that $\Phi(c)$ lands in this fiber.

There is a second and different way that the various containers can overlap, so we show another picture. Figure 12.2 shows the fiber over $(x, 2/5)$ where $x \approx -1/3$. This point lies in a different isosceles triangle in the base partition. The remaining pictures look like either Figure 12.2 or Figure 12.3.

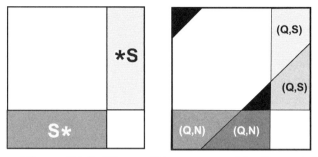

Figure 12.2: The partition compared to the container.

In Figure 12.3, we use 4 images. The top figures (repeated) show the PET picture. This part looks like Figure 12.2. All 4 containers make their appearance in Figure 12.3. The bottom left side of Figure 12.3 shows the containers for (P, N) and (P, S). The bottom right side of Figure 12.3 shows the containers for (Q, N) and (Q, S). Some of the containers from the left overlap with the containers from the right, as is indicated by some crosshatching and shading. Once again, the region covered exactly once is the union of $S*$ and $*S$, and the directions work out perfectly.

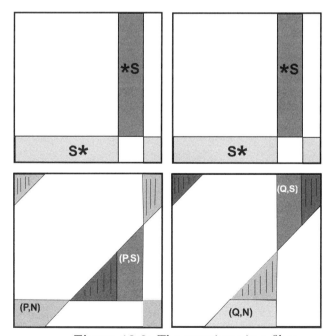

Figure 12.3: The containers in a fiber.

In all cases there is a perfect match-up between the two schemes. As in the vertical case, both schemes are combinatorially stable. Over the same open isosceles triangle in the base partition the combinatorics of the partition does not change. This reduces the check to 18 fibers, as in the vertical case. Again, we make the check by direct inspection. The alternate check discussed above also works.

This completes the proof of Theorem 8.2.

Computer Tie-In: You can make this check for yourself. Follow all the steps for the vertical case but use the setting *slice: hor* and use the types corresponding to horizontal light points.

Part 3. The Graph PET

Chapter Thirteen

Graph Master Picture Theorem

13.1 CHAPTER OVERVIEW

This chapter starts Part 3 of the monograph. The purpose of this chapter is prove prove Theorem 0.4, the Graph Master Picture Theorem. We will prove Theorem 0.4 in two different ways. This chapter contains the first proof. The proof here deduces Theorem 0.4 from Theorem 13.2, which is a restatement of [**S1**, Master Picture Theorem] with minor cosmetic changes. Subsequent chapters will give a self-contained proof of Theorem 16.9, a more general result which, for each individual parameter, specializes to Theorem 0.4 in the case of special orbits on a kite.

In §13.2 we discuss the special outer billiards orbits on kites. In §13.3 we define the arithmetic graph, which is an arithmetical way of encoding the behavior of a certain first return map of the special orbits. In §13.4 we state Theorem 13.2, a slightly modified and simplified version of [**S1**, Master Picture Theorem]. In §13.5 we deduce Theorem 0.4 from Theorem 13.2 and one extra piece of information. Finally, in §13.6 we list the polytopes comprising the partition associated to Theorems 13.2 and 0.4.

13.2 SPECIAL ORBITS

We will describe the situation when $A = p/q \in (0, 1)$ is rational. We need not take pq even here. Let $\mathbf{Z}_1$ denote the set of odd integers. We consider outer billiards on the kite K_A from Figure 0.1. Again, the verties of K_A are

$$(-1, 0), \quad (0, 1), \quad (0, -1), \quad (A, 0). \tag{13.1}$$

The *special orbits* are the orbits which lie on the set of horizontal lines having odd integer Y-intercepts. We say that a *special interval* is an open horizontal interval of length $2/q$ centered on a point $(a/q, b)$ with $a, b \in \mathbf{Z}_1$. Let θ_A denote the second iterate of the outer billiards map. This map is a piecewise translation. Recall that $\Xi = \mathbf{R}_+ \times \{-1, 1\}$.

Lemma 13.1 *The following is true.*

- θ_A *is defined on any special interval I, and $\theta_A(I)$ is another special interval. That is θ_A permutes the special intervals.*

- *For any special interval I, there are integers $m < 0 < n$ such that $\theta_A^m(I), \theta_A^n(I) \subset \Xi$.*

Proof: The first statement is [**S1**, Lemma 2.1]. Sketch: Reflection in the vertices of K_A preserves $\boldsymbol{R} \times \boldsymbol{Z}_1$. Second, when we extend the sides of of K_A they intersect $\boldsymbol{R} \times \boldsymbol{Z}_1$ in points which are the endpoints of special intervals. Since θ_A fails to be defined only on such lines, θ_A is defined on the special intervals and permutes them.

The second statement is essentially [**S1**, Lemma 2.3]. The hypotheses in [**S1**, Lemma 2.3] state that I should start out on Ξ, but there is nothing in the proof that requires that. Indeed, the proof amounts to showing that after a special interval leaves Ξ under the application of powers of θ_A, the image returns later on.

Here is a rough sketch of the proof of [**S1**, Lemma 2.3]. Let $\{a_n\}$ be the θ_A-orbit of a point a_0 on a special interval. The difference $a_{n+1} - a_n$ is $2\nu_n$, where ν_n is some vector which points from one vertex of K_A to another. Far away from K_A there are long stretches where the function $n \to \nu_n$ is constant. The changes occur when a_n comes close to certain rays extending the sides of K_A. An easy geometric analysis of the situation shows that the sequence $a_n, a_{n+1}, a_{n+1}, ...$ generally circulates around K_A, making what looks roughly like a giant octagon. (See §14.2.) On every revolution these points hit Ξ. The same is true if we consider negative powers of θ_A. The situation when the orbit stays close to K_A is handled with a case-by-case analysis. $\square$

13.3 THE ARITHMETIC GRAPH

Up to a shift in the indexing of the orbit, every special orbit is combinatorially identical to the orbit of a point of the form

$$\left(2mA + 2n + \frac{1}{q}, \pm 1\right), \qquad m, n \in \boldsymbol{Z}, \qquad 2mA + 2n + \frac{1}{q} > 0. \quad (13.2)$$

These points are the centers of the special intervals in Ξ, and the orbit of any point on a special orbit has the same combinatorial structure as the orbit of the midpoint.

Let $\Theta_A : \Xi \to \Xi$ denote the first return map of θ_A to Ξ. Suppose that m_0, n_0 are integers such that $2m_0A + 2n_0 \geq 0$. Then, by definition, there are integers m_1, n_1, with $2m_1A + 2n_1 \geq 0$ such that

$$\Theta_A\left(2m_0A + 2n_0 + \frac{1}{q}, 1\right) = \left(2m_1A + 2n_1 + \frac{1}{q}, \epsilon\right). \quad (13.3)$$

We prove in [**S1**, §2.4] that in fact $\epsilon = (-1)^{m_0+m_1+n_0+n_1}$.

We form a directed graph Γ_A whose vertices are $\boldsymbol{Z}^2$ by joining (m_0, n_0) to (m_1, n_1) by a directed edge if and only if these points are related as in Equation 13.3 and these points are less than $(p+q)/2$ apart. We make this last stipulation because in the rational case there are infinitely many choices

of (m'_1, n'_1) which are so related to (m_0, n_0) and we want to pick the nearest one. We proved in [S1] that the differences $m_0 - m_1$ and $n_0 - n_1$ lie in $\{-1, 0, 1\}$. Our proof of the Quasi-Isomorphism Theorem will give a second proof of this fact. We call Γ_A the *arithmetic graph*.

Remark: We are following the treatment in [S1], but in hindsight it is more natural to consider the second return map of θ to the union of lines $\widehat{\Xi} = \boldsymbol{R} \times \{-1, 1\}$. We also call this map Θ. The restriction of Θ to Ξ agrees with the old definition of Θ, and both sets $\boldsymbol{R}_\pm \times \{-1, 1\}$ are invariant sets for the new extended map. Making this extension allows us to extend Γ_A to the subset of $\boldsymbol{Z}^2$ below the line L of slope $-A$ through the origin. Indeed, there is a canonical bijection between the components of Γ_A above L and the components of Γ_A below L: Components corresponding to the same orbit of Θ are paired up.

We will describe the situation when pq is even. When pq is odd the situation is similar but differs in fine details. We only care about the case when pq is even. Let Γ_A^+ denote the portion of Γ_A above the line L. Let

$$\tau(m, n) = (m, n) + 2(q, -p). \tag{13.4}$$

The map

$$f(m, n) = \left(2Am + 2n + \frac{1}{q}, (-1)^{m+n+1}\right) \tag{13.5}$$

sets up a bijection from $\boldsymbol{Z}^2/\tau$ to the set of centers of the special intervals on Ξ. The bijection intertwines the curve-following dynamics on Γ_A^*/τ with the first return map to Ξ. In particular, f sets up a bijection between the components of Γ_A^+/τ and the special orbits in such a way (after suitable indexing) that the kth vertex of an arithmetic graph polygon γ corresponds to the kth point of $O_\gamma \cap \Xi$. Here O_γ is the special orbit corresponding to γ.

Computer Tie-In: On the main program, open up the *planar* window. Turn off all the features on the *plaid model* control panel. Open up the *arithmetic graph* control panel and turn on the *polygons* feature. In the planar window you will see the image $T_A(\Gamma_A)$ where T_A is the affine transformation from the Quasi-Isomorphism Theorem.

13.4 A PRELIMINARY RESULT

We fix $A = p/q \in (0, 1)$ as above. In this section we need not take pq even. As in the case of the plaid PET, we work in the space $\widehat{Y} = \boldsymbol{R}^3 \times [0, 1]$. This time, we use the coordinates (x, y, z, A) on $\widehat{Y}$. The only difference is that we are calling the 4th coordinate A rather than P. The slices of the respective PETs turn out to be related by the equation $P = 2A/(1 + A)$, an equation we have encountered many times already.

Let Λ' denote the abelian group generated by the following affine transformations.

- $T_1(x, y, z, A) = (x+1, y-1, z-1, A)$.

- $T_2(x, y, z, A) = (x, y+1+A, z+1-A, A)$.

- $T_3(x, y, z, A) = (x, y, z+1+A, A)$.

We introduce a map $\Psi''_A : \mathbf{Z}^2 \to \widehat{Y}$.

$$\Psi''_A(m, n) = (s/2, s/2, s/2, A), \qquad s = 2Am + 2n + \frac{1}{q}. \qquad (13.6)$$

We have left in the factor of 2 to better show the connection to Equation 13.3.

Here is the Master Picture Theorem from [**S1**, §6].

Theorem 13.2 *There are two Λ'-invariant partitions of $\widehat{Y}$ into convex integral polytopes. Each polytope in each partition is labeled by a pair of integers $(i, j) \in \{-1, 0, 1\}$. The image $\Psi''_A(c)$ lies in the interior of one such polygon for each $c \in \mathbf{Z}^2$ and each rational parameter $A = p/q \in (0, 1)$. If $\Psi''_A(c)$ lies in a polytope in the first partition with label (i_1, j_1) and a polytope in the second partition with label (i_2, j_2), then the arithmetic graph has the edge connecting c to $c + (i_1, j_1)$ and the edge connecting c to $c + (i_2, j_2)$.*

We call the two partitions from Theorem 13.2 the $(+)$ graph partition and the $(-)$ graph partition. We give a detailed description of these partitions in [**S1**, §6]. In §13.6 we list the 14 fundamental polytopes of the $(+)$ partition, together with their labels. We will also explain how to get the $(-)$ partition from the $(+)$ partition. As with the Plaid Master Picture Theorem, our listing, together with the formula for Ψ''_A, gives a complete statement of Theorem 13.2.

Discrepancies: There are 3 small differences between our presentation of the Master Picture Theorem here and the one given in [**S1**]. We mention these differences here.

First, we have switched the first and third coordinates. Thus, the vertex (x, y, z, A) of a polytope here corresponds to the vertex (z, y, x, A) in [**S1**]. This change makes the Master Picture Theorem line up more gracefully with the corresponding result for the plaid model. This switch also effects the definition of the lattice Λ'.

Second, here we have one classifying map and two partitions, whereas in [**S1**] we have two classifying maps, differing from each other by translations, and two slightly different partitions. The two partitions in [**S1**] are translates of the ones here respectively by the vectors $(0, 1, 0, 0)$ and $(-1, 0, 0, 0)$.

Third, in [**S1**] the map in Equation 13.3 does not use the "offset" $1/q$ but rather an infinitesimally small positive number ι. In fact, any value in $(0, 2/q)$ would yield the same results.

13.5 THE PET STRUCTURE

Now we explain how to deduce the Graph Master Picture Theorem, Theorem 0.4, from Theorem 13.2 and one additional piece of information.

Orientation Criterion: For each $\zeta \in \mathbf{Z}^2$, there are two directed edges of Γ_A incident to ζ. One points away from ζ and one points towards ζ. One of the two partition pieces containing $\Psi''_A(\zeta)$ determines the away-pointing edge and the other determines the towards-pointing edge. Tracing through [**S1**, Theorem 6.1] we get the following recipe. The $(+)$ partition determines the away-pointing edge at $\zeta = (m, n)$ if and only if the following quantity is even:

$$\rho_A(\zeta) = \text{floor}(s), \qquad s = (A+1)m + \frac{1}{2q}. \qquad (13.7)$$

The Space: Here we define the space Y and the first partition of Y in the Graph Master Picture Theorem. Let $\Lambda \subset \Lambda'$ denote the group generated by T_1^2, T_2 and T_3. Let

$$Y = \widehat{Y}/\Lambda. \qquad (13.8)$$

This is the space in the Graph Master Picture Theorem, Theorem 0.4.

The First Partition: The polytopes in the $(+)$ partition are such that all their first coordinates are 0 or 1. Let U_+ be the union of these polytopes. Likewise define U_-. The union

$$\bigcup_{T \in \langle T_2, T_3 \rangle} T(U_+)$$

completely fills up the slab $[0, 1] \times \mathbf{R}^2 \times [0, 1]$. The same goes when we replace U_+ with U_-. We get Λ-invariant partition by taking

$$\bigcup_{T \in \Lambda} T(U_+ \cup T_1(U_-)). \qquad (13.9)$$

When $k \in \mathbf{Z}$, the slabs $[k, k+1] \times \mathbf{R}^2 \times [0, 1]$ are alternately filled with pieces from the $(+)$ partition and pieces from the $(-)$ partition. We call these regions *(+) slabs* and *(-) slabs* accordingly. When we quotient out by Λ we get a partition of Y.

Intertwining Map: Define

$$\Psi'_A(x, y) = (t + t', t - t', t - t') \bmod \Lambda \qquad t = Ax + y + \frac{1}{2q}, \qquad t' = x - y.$$
$$(13.10)$$

This is the map in Theorem 0.4. Note that Ψ''_A and Ψ'_A agree mod Λ'.

Given $\zeta \in \mathbf{Z}^2$, the image $\Psi'_A(\zeta)$ lies in a $(+)$ slab when $\rho_A(\zeta)$ is even and $\Psi'_A(\zeta)$ lies in a $(-)$ slab when $\rho_A(\zeta)$ is odd. Therefore, the labels in the partition here determine the away-pointing edge of Γ_A incident to ζ.

Second PET Partition: The second partition of Y is obtained by moving the polytopes in the first partition according to their labels, as follows: Suppose that the polytope P has the label (i,j) and vertices $V_1, ..., V_m$. We let P^* be the new polytope whose vertices are $V_1^*, ..., V_m^*$ where

$$V_k^* = V_k + (d_k, -d_k, -d_k, 0), \qquad d_k = iV_{k4} + j. \qquad (13.11)$$

Here v_{k4} is the 4th coordinate of V_k.

Equation 13.11 is designed so that this operation is compatible with the curve-following dynamics on Γ_A. If $\zeta \to \zeta'$ and $\zeta^* = \zeta + (i,j)$, then

$$\Psi_A'(\zeta^*) - \Psi_A'(\zeta) = (d_k, -d_k, -d_k, 0).$$

It follows from Theorem 13.2 and the orientation criterion that the second union of polytopes is indeed a partition of Y. Each polytope in the first partition is carried to its counterpart in the second partition by an affine map and every polytope in sight, when lifted to $\widehat{Y}$, is integral. Thus, our PET is a fibered integral PET.

Computer Tie-In: Open the *graph PET* window. The *graph PET* shows your choice of the forward and backward partition. The *tiling extent* controls allow you to either plot the tiles in a single fundamental domain or plot a larger portion of the tiles. Drag the mouse on the *graph PET slicer* window to explore the partitions. If you open up the *polytope info* window and then click on a polytope in the *graph PET window*, the *polytope info* window will show the vertices of the polytope, the label associated to the polytope, and also a segment that indicates the edge of $T_A(\Gamma_A)$ that is determined by the label. Here T_A is the affine transform from the Quasi-Isomorphism Theorem mentioned in the introduction.

To see the Graph Master Picture Theorem in action, follow the instructions for the previous computer tie-in, so that you are viewing $T_A(\Gamma_A)$ for the selected parameter. Turn on the *nearest grid* feature on the *arithmetic graph* control panel. You can select a vertex of the *planar* window using the middle mouse button or key-x. The program only allows you to select the centers of unit integer squares. However, the program will highlight the nearest point in the graph grid. Soemtimes the point is highlighted with a solid orange dot and sometimes with an orange circle. The choice is determined by a parity consideration.

The *graph PET* window shows the image of this highlighted point under the map $\Psi_A = \Psi_A' \circ T_A^{-1}$. (We will compute Ψ_A in §18.2. This map is nicer than Ψ_A'.) You can see how the geometric segment in the *polytope info* window matches the local picture of $T_A(\Gamma_A)$.

13.6 THE FUNDAMENTAL POLYTOPES

Each partition is the union of 28 polytopes. We just list the polytopes of the (+) partition. Each polytope in the (−) partition is obtained from the polytope in the (+) partition by applying the involution

$$(x, y, z, A) \rightarrow (1, A, 2 + A, A) - (x, y, z, 0). \tag{13.12}$$

The label $(-i, -j)$ of the kth polytope in the (−) partition is the negative of the label (i, j) of the kth polytope in the (+) partition.

$$\begin{bmatrix} 0 \\ -1 \\ 0 \\ 0 \end{bmatrix} \begin{bmatrix} 0 \\ 0 \\ 0 \\ +1 \end{bmatrix} \begin{bmatrix} 0 \\ -1 \\ +1 \\ +1 \end{bmatrix} \begin{bmatrix} 0 \\ 0 \\ +1 \\ +1 \end{bmatrix} \begin{bmatrix} +1 \\ 0 \\ +1 \\ +1 \end{bmatrix} \quad (0, +1)$$

$$\begin{bmatrix} 0 \\ -1 \\ 0 \\ +1 \end{bmatrix} \begin{bmatrix} +1 \\ -1 \\ 0 \\ 0 \end{bmatrix} \begin{bmatrix} +1 \\ -1 \\ 0 \\ +1 \end{bmatrix} \begin{bmatrix} +1 \\ 0 \\ 0 \\ +1 \end{bmatrix} \begin{bmatrix} +1 \\ -1 \\ +1 \\ +1 \end{bmatrix} \quad (0, +1)$$

$$\begin{bmatrix} 0 \\ 0 \\ +1 \\ 0 \end{bmatrix} \begin{bmatrix} 0 \\ +1 \\ +1 \\ +1 \end{bmatrix} \begin{bmatrix} 0 \\ 0 \\ 2 \\ +1 \end{bmatrix} \begin{bmatrix} 0 \\ +1 \\ 2 \\ +1 \end{bmatrix} \begin{bmatrix} +1 \\ +1 \\ 2 \\ +1 \end{bmatrix} \quad (-1, 0)$$

$$\begin{bmatrix} 0 \\ 0 \\ 0 \\ 0 \end{bmatrix} \begin{bmatrix} 0 \\ 0 \\ +1 \\ +1 \end{bmatrix} \begin{bmatrix} +1 \\ 0 \\ +1 \\ +1 \end{bmatrix} \begin{bmatrix} +1 \\ +1 \\ +1 \\ +1 \end{bmatrix} \begin{bmatrix} +1 \\ 0 \\ 2 \\ +1 \end{bmatrix} \quad (-1, 0)$$

$$\begin{bmatrix} 0 \\ -1 \\ +1 \\ +1 \end{bmatrix} \begin{bmatrix} +1 \\ -1 \\ +1 \\ 0 \end{bmatrix} \begin{bmatrix} +1 \\ -1 \\ +1 \\ +1 \end{bmatrix} \begin{bmatrix} +1 \\ 0 \\ +1 \\ +1 \end{bmatrix} \begin{bmatrix} +1 \\ -1 \\ 2 \\ +1 \end{bmatrix} \quad (+1, 0)$$

$$\begin{bmatrix} 0 \\ 0 \\ 0 \\ 0 \end{bmatrix} \begin{bmatrix} +1 \\ 0 \\ 0 \\ 0 \end{bmatrix} \begin{bmatrix} +1 \\ 0 \\ +1 \\ 0 \end{bmatrix} \begin{bmatrix} +1 \\ 0 \\ +1 \\ +1 \end{bmatrix} \begin{bmatrix} +1 \\ +1 \\ +1 \\ +1 \end{bmatrix} \begin{bmatrix} +1 \\ 0 \\ 2 \\ +1 \end{bmatrix} \quad (-1, -1)$$

$$\begin{bmatrix} 0 \\ -1 \\ 0 \\ 0 \end{bmatrix} \begin{bmatrix} 0 \\ 0 \\ 0 \\ 0 \end{bmatrix} \begin{bmatrix} +1 \\ 0 \\ 0 \\ 0 \end{bmatrix} \begin{bmatrix} +1 \\ +1 \\ 0 \\ +1 \end{bmatrix} \begin{bmatrix} +1 \\ 0 \\ +1 \\ +1 \end{bmatrix} \begin{bmatrix} +1 \\ +1 \\ +1 \\ +1 \end{bmatrix} \quad (-1, 0)$$

$$\begin{bmatrix}0\\-1\\+1\\0\end{bmatrix}\begin{bmatrix}0\\0\\+1\\0\end{bmatrix}\begin{bmatrix}0\\0\\+1\\+1\end{bmatrix}\begin{bmatrix}+1\\0\\+1\\0\end{bmatrix}\begin{bmatrix}0\\-1\\2\\+1\end{bmatrix}\begin{bmatrix}0\\0\\2\\+1\end{bmatrix}\begin{bmatrix}+1\\0\\2\\+1\end{bmatrix}\quad(+1,0)$$

$$\begin{bmatrix}0\\-1\\0\\0\end{bmatrix}\begin{bmatrix}0\\0\\0\\0\end{bmatrix}\begin{bmatrix}0\\0\\0\\+1\end{bmatrix}\begin{bmatrix}0\\+1\\0\\+1\end{bmatrix}\begin{bmatrix}+1\\+1\\0\\+1\end{bmatrix}\begin{bmatrix}0\\0\\+1\\+1\end{bmatrix}\begin{bmatrix}+1\\0\\+1\\+1\end{bmatrix}\begin{bmatrix}+1\\+1\\+1\\+1\end{bmatrix}\quad(-1,+1)$$

$$\begin{bmatrix}0\\0\\0\\0\end{bmatrix}\begin{bmatrix}0\\+1\\0\\+1\end{bmatrix}\begin{bmatrix}0\\-1\\+1\\0\end{bmatrix}\begin{bmatrix}0\\0\\+1\\0\end{bmatrix}\begin{bmatrix}0\\0\\+1\\+1\end{bmatrix}\begin{bmatrix}+1\\0\\+1\\0\end{bmatrix}\begin{bmatrix}0\\+1\\+1\\+1\end{bmatrix}\begin{bmatrix}+1\\+1\\+1\\+1\end{bmatrix}\quad(0,+1)$$

$$\begin{bmatrix}0\\-1\\0\\0\end{bmatrix}\begin{bmatrix}+1\\-1\\0\\0\end{bmatrix}\begin{bmatrix}0\\0\\0\\+1\end{bmatrix}\begin{bmatrix}+1\\0\\0\\0\end{bmatrix}\begin{bmatrix}+1\\0\\0\\+1\end{bmatrix}\begin{bmatrix}+1\\+1\\0\\+1\end{bmatrix}\begin{bmatrix}+1\\-1\\+1\\0\end{bmatrix}\begin{bmatrix}+1\\0\\+1\\+1\end{bmatrix}\quad(0,+1)$$

$$\begin{bmatrix}0\\0\\+1\\0\end{bmatrix}\begin{bmatrix}0\\0\\+1\\+1\end{bmatrix}\begin{bmatrix}+1\\0\\+1\\0\end{bmatrix}\begin{bmatrix}0\\+1\\+1\\+1\end{bmatrix}\begin{bmatrix}+1\\+1\\+1\\+1\end{bmatrix}\begin{bmatrix}0\\0\\2\\+1\end{bmatrix}\begin{bmatrix}+1\\0\\2\\+1\end{bmatrix}\begin{bmatrix}+1\\+1\\2\\+1\end{bmatrix}\quad(0,0)$$

$$\begin{bmatrix}0\\-1\\0\\0\end{bmatrix}\begin{bmatrix}0\\-1\\0\\+1\end{bmatrix}\begin{bmatrix}+1\\-1\\0\\0\end{bmatrix}\begin{bmatrix}0\\0\\0\\+1\end{bmatrix}\begin{bmatrix}+1\\0\\0\\+1\end{bmatrix}\begin{bmatrix}0\\-1\\+1\\+1\end{bmatrix}\begin{bmatrix}+1\\-1\\+1\\0\end{bmatrix}\begin{bmatrix}+1\\-1\\+1\\+1\end{bmatrix}\begin{bmatrix}+1\\0\\+1\\+1\end{bmatrix}\quad(+1,+1)$$

$$\begin{bmatrix}0\\-1\\0\\0\end{bmatrix}\begin{bmatrix}0\\0\\0\\0\end{bmatrix}\begin{bmatrix}+1\\0\\0\\0\end{bmatrix}\begin{bmatrix}0\\-1\\+1\\0\end{bmatrix}\begin{bmatrix}0\\-1\\+1\\+1\end{bmatrix}\begin{bmatrix}+1\\-1\\+1\\0\end{bmatrix}$$

$$\begin{bmatrix}0\\0\\+1\\+1\end{bmatrix}\begin{bmatrix}+1\\0\\+1\\0\end{bmatrix}\begin{bmatrix}+1\\0\\+1\\+1\end{bmatrix}\begin{bmatrix}0\\-1\\2\\+1\end{bmatrix}\begin{bmatrix}+1\\-1\\2\\+1\end{bmatrix}\begin{bmatrix}+1\\0\\2\\+1\end{bmatrix}\quad(0,0)$$

Chapter Fourteen

Pinwheels and Quarter Turns

14.1 CHAPTER OVERVIEW

In the next three chapters, we will prove a generalization of the Graph Master Picture Theorem which works for any convex polygon P without parallel sides. Our final result is Theorem 16.9, though Theorems 15.1 and 16.1 are even more general. None of this material is needed for the rest of the monograph.

We fix P for the remainder of Part 3 of the monograph. Let θ denote the second iterate of the outer billiards map defined on $\mathbf{R}^2 - P$. In §14.2 we will generalize a construction in [**S1**] and define a map closely related to θ, which we call the *pinwheel map*. In §14.3 we will see that, for the purposes of studying unbounded orbits, the pinwheel map carries all the information contained in θ. However, in general the pinwheel map ignores "a bounded amount" of information contained in θ. In the case of kites, the pinwheel map contains all the information. The general version of the Graph Master Picture Theorem relates to the pinwheel map.

In §14.4 we will define another dynamical system called a *quarter turn composition*. A QTC is a certain kind of piecewise affine map of the infinite strip $\mathbf{S}$ of width 1 centered on the X-axis. In §14.5 we will see that the pinwheel map naturally gives rise to a QTC and indeed the pinwheel map and the QTC are conjugate. In §14.6 we will explain how this all works for kites.

14.2 THE PINWIIEEL MAP

Here we recall some work we did in [**S2**]. Let Σ be an infinite strip in the plane and let V be a vector that spans Σ in the sense that the tail of V lies on one component of $\partial\Sigma$ and the head of V lies on the other component. See Figure 14.1.

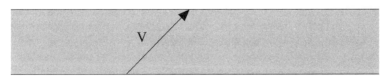

Figure 14.1: The vector V spans the strip.

The pair (Σ, V) defines a map $\tau : \boldsymbol{R}^2 \to \Sigma$, as follows.

$$\tau(p) = p + nV \tag{14.1}$$

Here $n \in \boldsymbol{Z}$ is the integer such that $p + nV \in \Sigma$. The map τ is well defined in the complement of a discrete infinite family of lines which are parallel to Σ. This family of lines contains the two lines of $\partial \Sigma$.

Let P and θ be as above.

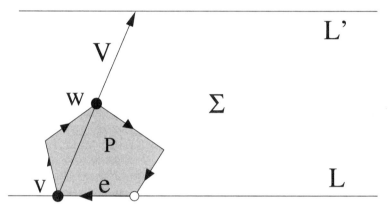

Figure 14.2: The strip associated to e.

We orient the edges of P clockwise. Given an edge e of P, we let L be the line extending e and we let L' be the line parallel to L so that the vertex w of P that lies farthest from L is equidistant from L and L'. We associate to e the pair (Σ, V), where Σ is the strip bounded by L and L', and $V = 2(w-v)$. See Figure 14.2.

We order the strips according to their slopes, so that one turns counterclockwise when changing from Σ_i to Σ_{i+1}. This ordering typically does not coincide with the cyclic ordering on the edges. The corresponding composition

$$T_1 = \tau_n \circ \dots \circ \tau_1 \tag{14.2}$$

is what we call the *pinwheel map*. T_1 is a map from Σ_1 to Σ_1.

To describe the connection between T_1 and outer billiards, we first work outside some large compact subset $K \subset \boldsymbol{R}^2$. Suppose we start with a point $p_1 \in \Sigma_1$. Then $\theta^k(p_1) = p_1 + kV_2$ for $k = 1, 2, 3, \dots$ This general rule continues until we reach an exponent k_1 such that $p_2 = \theta^{k_1}(p_1) \in \Sigma_2$. Then we have $\theta^k(p_2) = p_2 + kV_3$ for $k = 1, 2, 3, \dots$, until we reach an exponent k_2 such that $p_3 = \theta^{k_2}(p_2) \in \Sigma_3$. And so on. See Figure 1.4. We eventually reach a point $p_{n+1} \in \Sigma_1$, and the map $p_1 \to p_{n+1}$ is the first return map.

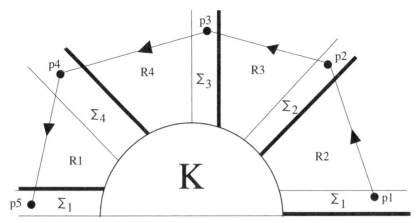

Figure 14.3: Far from the origin.

The connection between the first return map and θ, for orbits *far away* from the polygon, appears in almost every paper on polygonal outer billiards. However, for points which start out near P, the connection is much less clear.

From the connection between outer billiards and the pinwheel map, we see one essential feature of our strips and vectors. (This connection can also be seen directly in terms of the polygon.) We have already mentioned that V_i spans Σ_i. Were we to consider the map θ^{-1} in place of θ, we would produce the sequence of strips $(\Sigma_{j-1}, -V_j)$. Therefore V_j spans Σ_{j-1} as well as Σ_j. We can put this in another way. $\Sigma_{j-1} \cap \Sigma_j$ is a parallelogram, and $2V_j$ is always one of the diagonals of this parallelogram.

14.3 OUTER BILLIARDS AND THE PINWHEEL MAP

We call two maps $f, g : \Sigma_1 \to \Sigma_1$ *unboundedly equal* if there is a disk $K \subset \mathbf{R}^2$ with the following properties.

1. $\Sigma - K$ consists of 2 connected components.

2. $f = g$ on $\Sigma - K$.

3. If a and b are any two points in the same component of $\Sigma - K$ then b is in the forward (respectively backward) f-orbit of a if and only if b is in the forward (respectively backward) g-orbit of a.

We point out that one can construct examples where the first two statements are satisfied but not the third. (Hint: consider two foliations of the plane, one by spirals and one by concentric circles. Now squash these pictures into a strip.)

Here we describe the main result in [**S2**]. Let θ be the second iterate of the outer billiards map, as above.

Theorem 14.1 *Let T_1 be the pinwheel map on Σ_1 and let Θ_1 be the first return map of θ to Σ_1. The maps Θ_1 and T_1 are unboundedly equal on Σ_1.*

Remark: Theorem 14.2 tells us that we can replace the outer billiards system with the pinwheel map if we are only interested in the existence of unbounded orbits. However, for kites, we are interested in a more precise description of the outer billiards orbits. So, it might first appear that our switch to the pinwheel map causes us to lose some information about the orbits. However, for kites we have $\Theta_1 = T_1$ on the set $\Sigma_1 \cap S$ where Σ_1 is chosen to be the strip of slope -1 associated to the kite and S is the set of horizontal lines having odd integer Y-intercept. We will explain this in §14.6 below. The reader who wants to understand Theorem 14.1 for kites need not read [**S2**]. All that is needed is the fairly short [**S1**, Pinwheel Lemma].

Suppose we start at $a \in \Sigma_1$ and iterate the two maps Θ_1 and T_1. Let's say we produce points $c_1, c_2, \dots$ and $c_1', c_2', \dots$ As long as none of these points get close to the origin, we have $c_j = c_j'$ for all j and then we are simply saying that there is some index k so that $b = c_k = c_k'$. However, if the points move close to the origin then it might happen that $c_j \neq c_j'$. The orbits can diverge from each other. However, Theorem 14.1 is saying that these two sequences sync back up once they move far away from the origin. Here is a corollary of this result, also proved in [**S2**].

Theorem 14.2 *There is a canonical bijection between the set of unbounded orbits of θ and the set of unbounded orbits of T_1. The bijection is such that the θ-orbit O corresponds to the T_1-orbit which agrees with $O \cap \Sigma_1$ outside a compact set. In particular, outer billiards on P has unbounded orbits if and only if T_1 has unbounded orbits.*

14.4 QUARTER TURN COMPOSITIONS

In this section and the next we analyze the structure of the pinwheel map. Let **S** denote the strip of width 1 whose centerline is the X-axis.

Let $\square \subset \mathbf{S}$ be a rectangle with sides parallel to the coordinate axes. The top and bottom of $\square$ are supposed to lie in the top and bottom boundary of **S**. We define a *quarter turn* of $\square$ to be the order 4 affine automorphism of $\square$ which maps the right edge of R to the bottom edge of $\square$. This map essentially twirls $\square$ $1/4$ of a turn clockwise. For any $a > 0$ we distinguish 2 tilings of the strip **S** by $a \times 1$ rectangles. In *Tiling 0*, the origin is the center of a rectangle. In *Tiling 1*, the origin is the center of a vertical edge of a rectangle. For $q = 0, 1$ let $R_{q,a}$ denote the way which gives a quarter turn to each rectangle in Tiling q. The map $R_{q,a}$ is a piecewise affine automorphism of $\mathcal{S}$, defined everywhere except the vertical edges of the rectangles. We call $R_{q,a}$ a *quarter turn*.

We define the *shear*

$$S_s = \begin{bmatrix} 1 & -s \\ 0 & 1 \end{bmatrix}. \tag{14.3}$$

Here $s > 0$. The map S_s is a shear of $\mathbf{S}$ which fixes the centerline pointwise, moves points with positive y-coordinate backwards and points with negative y-coordinate forwards.

We define a *quarter turn composition* (QTC) to be a finite alternating composition $\mathcal{T}$ of quarter turns and shears. That is,

$$\mathcal{T} = S_{s_n} \circ R_{q_n, r_n} \circ \cdots \circ S_{s_1} \circ R_{q_1, r_1}. \tag{14.4}$$

- $q_1, ..., q_n \in \{0, 1\}$ specify the tiling offsets.

- $r_1, ..., r_n$ are the parameters for the widths of the rectangles.

- $s_1, ..., s_n > 0$ are the parameters for the shears.

We call n the *length* of the QTC.

It is convenient to define

$$\alpha_i = r_n / r_i. \tag{14.5}$$

The choice of n as a special index is arbitrary; any other choice leads to the same definitions.

Definitions: Here are two definitions that will come up later.

- We call $\mathcal{T}$ *quasi-rational* if $\alpha_i \in \mathbf{Q}$ for all i.

- We call $\mathcal{T}$ *finitary* if $\mathcal{T}$ is a piecewise translation, and the set

$$\{\mathcal{T}(p) - p| \ p \in \mathbf{S}\} \tag{14.6}$$

of possible translations is finite.

14.5 THE PINWHEEL MAP AS A QTC

Now we will recognize the pinwheel map T_1 as a QTC. We will actually consider both T_1 and T_1^2. Ultimately, we will like T_1^2 better because it is fairly close to the identity map. It corresponds to moving the point all the way around P rather than halfway around.

The Idea: Before we launch into the rather involved proof, we explain the basic idea behind the proof. The pinwheel map involves a composition of maps between different strips and a QTC involves a composition of maps all taking place on the same strip $\mathcal{S}$. The way we convert the pinwheel map to the QTC is that we identify all these other strips with $\mathbf{S}$ using carefully chosen affine transformations. The geometry of the polygon P gives a recipe for choosing these affine transformations well. As we discuss below, our computer programs check that we've got the formulas all correct for kites.

Now we get into the details. Let $(\Sigma_1, V_1), ..., (\Sigma_{2n}, V_{2n})$ denote the strip data above, repeated twice. Let $\tau_1, ..., \tau_{2n}$ be the corresponding strip maps. For the purpose of getting the signs right when we define certain maps, we normalize the picture by a suitable affine transformation. We assume that $\Sigma_1 = \mathbf{S}$ and that $\Sigma_2, ..., \Sigma_n$ all have positive slope. However, after we define our maps, we will not insist on this normalization. For $k = 1, ..., n$, we define map

$$A_{k,\pm} : \Sigma_k \to \mathbf{S} \tag{14.7}$$

by the following rules.

- $A_{k,\pm}$ is area and orientation preserving, and maps points with large positive x-coordinate to points with large positive x-coordinate.

- $A_{k,\pm}$ maps the parallelogram $\Sigma_k \cap \Sigma_{k\pm1}$ to a rectangle, and the head point of V_k to the origin.

Let $\rho(x, y) = (-x, -y)$ be reflection about the origin. We define

$$A_{n+k,\pm} = \rho \circ A_{k,\pm}, \tag{14.8}$$

$$R_k = A_{k+1,-} \circ \tau_k \circ A_{k,+}^{-1}, \qquad S_k = A_{k+1,+} \circ (A_{k+1,-})^{-1}. \tag{14.9}$$

Now that we have defined these maps, we drop the assumption about the slopes of the strips. In general, the maps are defined in such a way that the whole construction is natural under affine conjugation.

Lemma 14.3 S_k *is an affine shear, as in Equation 14.3.*

Proof: The maps $A_{k+1,\pm}$ are both area preserving, orientation preserving, sense preserving affine bijections from Σ_{k+1} to $\mathbf{S}$, and they both map the same point to the origin. From this description, it is clear that S_k has the equation given in Equation 14.3. The only thing that remains to prove is that $b > 0$. That is, S_k shears points in $\mathcal{S}$ with positive y-coordinate to the left.

To understand what is going on, we take $k = 1$. Now we normalize so that Σ_1 is horizontal and Σ_2 is vertical, and the positive senses of these strips go along the positive coordinate axes. See Figure 14.4. This situation forces Σ_3 to have negative slope. $\Sigma_1 \cap \Sigma_2$ is the thickly drawn square and $\Sigma_2 \cap \Sigma_3$ is the shaded parallelogram. The significant feature here is that the left side of the shaded parallelogram lies above the right side. This fact translates into the statement that $b > 0$ in Equation 14.3. $\square$

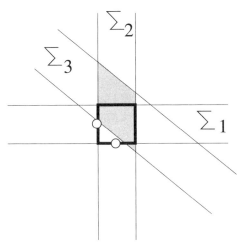

Figure 14.4: Placement of the strips.

Lemma 14.4 R_k *is a quarter turn map.*

Proof: Our proof will also identify the parameters of R_k. The parameter r_k is just the area of $\Sigma_k \cap \Sigma_{k-1}$. We will consider the case $k = 1$. We first discuss how our construction interacts with affine transformations. Let Δ be an affine transformation, which expands areas by δ. Let R_1' be the map associated to $\Delta(P)$. We have

$$R_1' = D_\delta \circ R_1 \circ D_\delta^{-1}, \qquad D_\delta(x,y) = (\delta x, y).$$

Thanks to this equation, it suffices to prove our result for any affine image of P. We normalize by an affine transformation so that

$$\Sigma_1 = \mathbf{R} \times [-1/2, 1/2], \qquad \Sigma_2 = [-1/2, 1/2] \times \mathbf{R}, \qquad V_2 = (-1, 1). \tag{14.10}$$

In this case, $A_{1,+}$ is the identity and $A_{2,-}$ is the clockwise order 4 rotation about the origin. Figure 14.5 shows the action of T_2 on Σ_1.

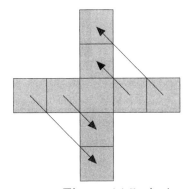

Figure 14.5: Action of τ_2.

From Figure 14.5, and from the description of $A_{1,+}$ and $A_{2,-}$, we see that there is a tiling $\mathcal{S}$ of $\mathbf{S}$ by unit squares and R_1 gives a clockwise quarter turn to each unit square. To finish the proof, we just have to see that $\mathcal{S}$ is one of the two special tilings discussed in §14.4.

Let e_j be the edge of P that lies in $\partial \Sigma_j$. Let $|V_j|$ be the segment underlying the vector V_j. One basic principle we use in our analysis is that $e_1, |V_1|, |V_2|$ make the edges of a triangle. Call this the *triangle property*.

Let c_j be the head of V_j. We have

$$R_{1,+}(c_1) = (0,0) = R_{2,-}(c_2). \tag{14.11}$$

There are two cases to consider. Suppose that c_1 is not incident to e_2. By the triangle property, c_1 is incident to V_2. Hence either $c_1 = c_2$ or c_1 is the tail vertex of V_2. But the tail vertex of V_2 is incident to e_2. This proves that $c_1 = c_2$. This situation implies that the common point $c = c_1 = c_2$ is the center of $\Sigma_1 \cap \Sigma_2$. From this information, and Equation 14.11, we conclude that $\mathcal{S}$ is Tiling 1. Suppose that c_1 is incident to e_2. Then c_1 lies on the centerline of Σ_1 and on the boundary of Σ_2. Hence c_1 is the midpoint of an edge of $\Sigma_1 \cap \Sigma_2$. Hence, the origin is the midpoint of an edge of a tile in $\mathcal{S}$. Hence $\mathcal{S}$ is Tiling 2. $\square$

We define

$$\mathcal{T}_P = S_n \circ R_n \circ \cdots \circ S_1 \circ R_1. \tag{14.12}$$

By construction, $\mathcal{T}_P$ is a QTC.

Let T_1 be the pinwheel map. By construction

$$T_1 = \rho \circ \mathcal{T}_P; \qquad T_1^2 = S_{2n} \circ R_{2n} \circ \cdots \circ S_1 \circ R_1. \tag{14.13}$$

Lemma 14.5 $\mathcal{T}_P^2$ *is finitary.*

Proof: We have $\mathcal{T}_P^2 = T_1^2$. The map T_1^2 is evidently a piecewise translation. We just need to prove that the set $\{T_1^2(p) - p \mid p \in S\}$ is finite. There is a sequence of numbers $m_1, ..., m_{2n}$ such that

$$T_1^2(p) - p = \sum_{i=1}^{2n} m_i V_i = \sum_{i=1}^{n} (m_i - m_{i+n}) V_i.$$

Here $V_1, ..., V_n$ are the vectors that arise in the strip maps, and we are setting $V_{i+n} = -V_i$. The number m_j refers to the analysis of Figure 14.3. Here m_j is the number of iterates of θ needed to carry the iterate lying in Σ_{j-1} to the iterate lying in Σ_j.

Far from the origin, the portion of the θ-orbit of p, going from p to $T_1^2(p)$, lies within a uniformly bounded distance of a centrally symmetric $2n$-gon. The point is that all the strips come within a uniform distance of the origin. From this property, we see that there is a uniform bound to $|m_i - m_{i+n}|$ for all i. Hence, there are only finitely many choices for $T_1^2(p) - p$. $\square$

14.6 THE CASE OF KITES

Now we explain the situation for the kite K_A. First of all, we reconcile the constructions in this chapter with the constructions in the last chapter. Let Σ_1 be the strip of slope -1 associated to K_A. Let S be the set of horizontal lines of odd integer slope. Recall that θ is the square of the outer billiards map and Θ is the second return map of θ to the set $\boldsymbol{R} \times \{-1, 1\}$.

Figure 14.6 shows how certain powers of θ give a natural map

$$\overline{\theta} : \Sigma \cap S \to \boldsymbol{R} \times \{-1, 1\}. \tag{14.14}$$

The map $\overline{\theta}$ just adds suitable integer multiples of the vector $(0, 4)$ to the components of $\Sigma_1 \cap S$. These vectors coincide with the action of a suitable power of θ. We have

$$\Theta = \overline{\theta} \circ \Theta_1^2 \circ \overline{\theta}^{-1}. \tag{14.15}$$

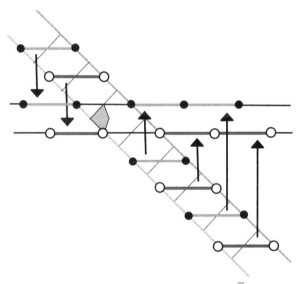

Figure 14.6: The map $\overline{\theta}$.

For kites, we deduce a strong version of Theorem 14.1 from [**S1**, Pinwheel Lemma].

Lemma 14.6 *Relative to the kite K_A the maps Θ_1 and T_1 agree on $\Sigma_1 \cap S$.*

Proof: [**S1**, Pinwheel Lemma] shows that $\overline{\theta} \circ T_1 \circ \overline{\theta}^{-1}$ and $\overline{\theta} \circ \Theta_1 \circ \overline{\theta}^{-1}$ coincide on $\boldsymbol{R}_+ \times \{-1, 1\}$. The proof actually shows that these two maps coincide on all of $\boldsymbol{R} \times \{-1, 1\}$. But $\overline{\theta}$ conjugates the two maps just considered to T_1 and Θ_1 respectively. Hence, these maps coincide on $\Sigma_1 \cap S$. $\square$

In view of Lemma 14.6 the map T_1^2, the square of the pinwheel map, is conjugate to the map Θ used to define the arithmetic graph in the last chapter. At the same time, T_1 is conjugate to the quarter turn composition $\mathcal{T}$ associated to K_A.

To describe $\mathcal{T}$ precisely, we first apply a similarity so that the strip Σ_1 of K is moved to the standard strip $\mathbf{S}$. Let $K' = h^{-1}(K)$ where

$$h(x, y) = (2x + 2y, -2x + 2y + 1). \tag{14.16}$$

See Figure 14.7 below.

Define

$$B = \frac{1 + A}{1 - A}, \qquad C = \frac{(1 + A)^2}{4A}. \tag{14.17}$$

The parameters associated to $\mathcal{T}$ are

$$q = (1, 1, 0, 1); \qquad r = (1, B, C, B); \qquad s = (B, BC, BC, B). \tag{14.18}$$

Under the identification of K and K', the set $\Sigma \cap S$ corresponds to the set Δ of thick black diagonal lines in Figure 14.7.

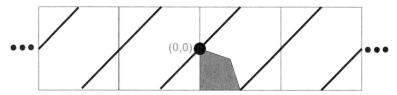

Figure 14.7: The set Δ contained in $\mathbf{S}$.

The map Θ is conjugate to $\mathcal{T}|_\Delta$. Thus, for the purposes of proving Theorems 0.4 and 13.2, the map $\mathcal{T}|_\Delta$ is equivalent to the map Θ.

Computer Tie-In: The subdirectory **QTCKite** contains an auxiliary program which illustrates the QTCs associated to kites. When you open this program, a control panel pops up and so does a window called *Quarter Turn Composition*. The initial parameter is $A = 2/9$ but you can change this if you like. If you click the middle mouse button or key-x over the *QTC* window, you will see the orbit of the maximal periodic tile containing the point you clicked. You can use the *range* control panel to coerce your clicks so that they lie in the set $\Delta \subset \mathbf{S}$ or you can allow your clicks to go anywhere in $\mathbf{S}$. If you use my program *Billiard King*, the companion program for [**S1**], you can see that the tiling you get from the QTC is the same as what is produced by outer billiards on the corresponding kite.

Chapter Fifteen

Quarter Turn Compositions and PETs

15.1 CHAPTER OVERVIEW

In this chapter we will prove a general compactification theorem for quarter turn compositions. Throughout the chapter we fix a length n quarter turn composition $\mathcal{T}$ as Equation 14.4. The domain for $\mathcal{T}$ is the infinite strip $\mathbf{S} = \mathbf{R} \times [-1/2, 1/2]$. Our domain for the compactification is the unit torus

$$\widehat{\mathbf{S}} = \mathbf{R}^{n+1}/\mathbf{Z}^{n+1}. \tag{15.1}$$

It is convenient to let $\mathbf{T}^d = \mathbf{R}^d/\mathbf{Z}^d$ denote the unit torus made from the first d-coordinates of $\mathbf{R}^{n+1}$. Thus $\widehat{\mathbf{S}} = \mathbf{T}^{n+1}$.

Theorem 15.1 *Suppose that $\mathcal{T}$ is a length n quarter turn composition. Then there is a locally affine map $\Psi : \mathbf{S} \to \widehat{\mathbf{S}}$ and an affine PET, $\widehat{\mathcal{T}} : \widehat{\mathbf{S}} \to \widehat{\mathbf{S}}$, such that $\Psi \circ \mathcal{T} = \widehat{\mathcal{T}} \circ \Psi$.*

1. *The map Ψ is injective if and only if $\mathcal{T}$ is not quasi-rational.*

2. *The closure of $\Psi(\mathbf{S})$ is a sub-torus of dimension $1 + d$, where d is the $\mathbf{Q}$-rank of $\mathbf{Q}(\alpha_1, ..., \alpha_{n-1})$.*

3. *If $\mathcal{T}^k$ is finitary, then the restriction of $(\widehat{\mathcal{T}})^k$ to the closure of $\Psi(\mathbf{S})$ is an ordinary PET.*

Here $\alpha_1, .., \alpha_{n-1}$ are as in Equation 14.5. The terms *finitary* and *quasi-rational* are defined at the end of §14.4.

In §15.2 we prove a well-known result from linear algebra which will help with the material in the following section. In §15.3 we define the map $\Psi : \mathbf{S} \to \widehat{\mathbf{S}}$ and study the dimension of its image as a function of the parameters of $\mathcal{T}$. Recall that $\mathcal{T}$ is a composition of shears and quarter turn maps. In §15.4 we will establish Lemma 15.6, which shows that Ψ interacts in the desired way with shears. In §15.5, we establish Lemma 15.7, which does the same thing for quarter turn maps. In §15.6 we combine Lemmas 15.6 and 15.7 to prove Theorem 15.1.

Computer Tie-In: Our auxiliary program in the directory **QTCKite** shows Theorem 15.1 in action for the case of kites. If you open this program, you can bring up an auxiliary popup window called the *PET* window. The

PET window shows the compactification in Theorem 15.1, though technically we modify it along the lines of Theorem 16.9 so the domain is an honest polytope rather than a torus.

Whenever you click on the QTC window, the corresponding periodic tile is plotted in the *PET* window (as well as the QTC window). The PET tile is 5-dimensional, as is the PET itself. The *PET* window plots various 2-dimensional slices. The *PET slicer* window lets you control which slice you see. If you press the *kite* button on the *PET slicer* window you can choose the slice to coincide with the image $\Psi(\mathbf{S})$. In this case, the tile in the *PET* window will look exactly the same as the tile in the QTC window. This feature is a powerful check that Theorem 15.1 really works, at least for kites.

15.2 A RESULT FROM LINEAR ALGEBRA

A codimension 1 subspace $V \subset \mathbf{R}^n$ is *rational* if V has a basis consisting of vectors in $\mathbf{Z}^n$. We define the projection $\pi : \mathbf{R}^n \to \mathbf{T}^n = \mathbf{R}^n / \mathbf{Z}^n$.

Lemma 15.2 *If $\pi(V)$ is not dense in $\mathbf{T}^n$ then V is rational.*

Proof: When $n = 1$ there is nothing to prove. When $n = 2$, a geodesic in $\mathbf{T}^2$ is closed if and only it has rational slope, and otherwise dense. Thus V contains a rational vector when $\pi(V)$ is not dense. Scaling, we see that V contains an integer vector. This proves the case $n = 2$.

The general case goes by induction. If $V = \mathbf{R}^{n-1}$ then we are done. So, assume $V \neq \mathbf{R}^{n-1}$. Let $V_k \subset V$ denote the subset consisting of vectors whose last coordinate is k. The union $\bigcup_{k \in \mathbf{Z}} \pi(V_k)$ is contained in $\mathbf{T}^{n-1}$ and cannot be dense there. In particular $\pi(V_0)$ is not dense in $\mathbf{T}^{n-1}$. But then V_0 has a basis of integer vectors.

Each $\pi(V_k)$ is a parallel translate of $\pi(V_0)$. Since the union of these translates is not dense in $\mathbf{T}^{n-1}$ there can be only finitely many of them. Hence there is some $N \neq 0$ such that $\pi(V_N) = \pi(V_0)$. Letting v_0 be any integer vector in V_0, we know that there is some vector v_N such that $\pi(v_N) = \pi(v_0)$. But then $v_N - v_0 \in \mathbf{Z}^n$. Hence $v_N \in \mathbf{Z}^n$. But then we can augment our basis for V_0 with v_N and this gives us an integer basis for V. $\square$

15.3 THE MAP

Let $\mathcal{T}$, $\mathbf{S}$ and $\widehat{\mathbf{S}}$ be as in Theorem 15.1. Let the $r_1, ..., r_n$ be the data for $\mathcal{T}$ as in §14.4. Recall that $\alpha_i = r_n / r_i$. We define $\Psi : \mathbf{S} \to \widehat{\mathbf{S}}$ by the formula

$$\Psi(x, y) = (\psi(x), [y]); \qquad \psi(x) = \left[\frac{x}{r_1}, ..., \frac{x}{r_n} \right]. \qquad (15.2)$$

Here $[p]$ denotes the image of the point p in the relevant space $(\boldsymbol{R}/\boldsymbol{Z})^k$. So, for instance, $[y] \in \boldsymbol{R}/\boldsymbol{Z}$.

Lemma 15.3 Ψ *is injective if and only if* $\mathcal{T}$ *is not quasi-rational.*

Proof: Ψ is not injective if and only if there are numbers x_1, x_2 such that $(x_1 - x_2)/r_i \in \boldsymbol{Z}$ for all i, which is true iff $\alpha_i = r_n/r_i \in \boldsymbol{Q}$ for all i. $\square$

Let d be the dimension of the $\boldsymbol{Q}$-vector space $\boldsymbol{Q}(\alpha_1, \ldots, \alpha_{n-1})$.

Lemma 15.4 *If* $d = n$ *then* $\Psi(\mathbf{S})$ *is dense in* $\widehat{\mathbf{S}}$.

Proof: It suffices to prove that $\psi(\boldsymbol{R})$ is dense in $\boldsymbol{T}^n$. Let $\pi : \boldsymbol{R}^n \to \boldsymbol{T}^n$ be the quotient map. Let X denote the closure of $\psi(R)$ in $\boldsymbol{T}^n$. Note that X is an abelian group. If $X \neq \boldsymbol{T}^n$, then X is a lower dimensional flat sub-torus of $\boldsymbol{T}^n$, and the connected component of $\pi^{-1}(X)$ through the origin is contained in a codimension 1 subspace $V \subset \boldsymbol{R}^n$ whose projection $\pi(V)$ is not dense in $\boldsymbol{T}^n$. By Lemma 15.2, the subspace V has an integer basis $v_1, \ldots, v_{n-1}$. But then

$$\det(v_1, \ldots, v_{n-1}, \psi(x)) = 0. \tag{15.3}$$

Dividing through by r_n this gives us a nontrivial rational relation amongst $\alpha_1, \ldots, \alpha_{n-1}, 1$. This is a contradiction. $\square$

Lemma 15.5 $\dim(\mathbf{S}^*) = d + 1$, *where* d *is the dimension of the* $\boldsymbol{Q}$-*vector space* $\boldsymbol{Q}(\alpha_1, \ldots, \alpha_{n-1})$.

Proof: Permuting the coordinates, it suffices to consider the case when $\alpha_{n-d+1}, \ldots, \alpha_{n-1}, 1$ are independent over $\boldsymbol{Q}$ and α_j is a rational combination of these last d variables for all $j \leq n - d$. Let $\pi : \boldsymbol{T}^n \to \boldsymbol{T}^d$ be projection onto the last d coordinates. By the previous result, $\pi(X) = \boldsymbol{T}^d$. To prove that $\dim(X) = d$ it suffices to prove that $X \cap \pi^{-1}(0, \ldots, 0)$ consists of finitely many points. Let p be a point in this intersection. We will show that the first coordinate of p can only take on finitely many values. The same argument works for the remaining coordinates.

We have some integer relation

$$c_1\alpha_1 = c_{n-d+1}\alpha_{n-d+1} + \ldots + c_{n-1}\alpha_{n-1} + c_n. \tag{15.4}$$

Multiplying through by r_n we have

$$\frac{c_1}{r_1} = \frac{c_{n+d-1}}{r_{n+d-1}} + \ldots + \frac{c_n}{r_n}. \tag{15.5}$$

Suppose $x \in \boldsymbol{R}$ is such that $\pi \circ \psi(x)$ is close to $(0, \ldots, 0)$. Then x/r_j is close to an integer for $j = n - d + 1, \ldots, n$. But then $c_j x/r_j$ is also close to an integer for $j = n - d + 1, \ldots, n$. But then $c_1 x/r_1$ is close to an integer. This argument shows that the first coordinate of any point of $F \cap \pi^{-1}(0, \ldots, 0)$ has the form $[k/c_1]$ for some $k \in \{1, \ldots, c_1\}$. In particular, this is a finite set of possibilities. $\square$

15.4 COMPACTIFYING SHEARS

In this section we deal with the shears comprising the quarter turn compositions. This is pretty easy.

Lemma 15.6 *Let $S(x, y) = (x - sy, y)$ as in Equation 14.3. There is an affine PET $\widehat{S} : \widehat{\mathbf{S}} \to \widehat{\mathbf{S}}$ such that $\Psi \circ S = \widehat{S} \circ \Psi$. Here*

$$\widehat{S}([x_1, ..., x_n, y]) = \left[x_1 - \frac{s}{r_1} y, ..., x_n - \frac{s}{r_n} y, y \right]. \tag{15.6}$$

$\widehat{S}$ is an affine PET, and $\mathbf{T}^n \times (-1/2, 1/2) \subset \widehat{\mathbf{S}}$ is an invariant domain for $\widehat{S}$. The linear part of $\widehat{S}$ is given by the matrix

$$\begin{bmatrix} 1 & 0 & 0 & \cdots & -s/r_1 \\ 0 & 1 & 0 & \cdots & -s/r_2 \\ 0 & 0 & 1 & \cdots & -s/r_3 \\ \cdots & & & & \\ 0 & 0 & 0 & \cdots & 1 \end{bmatrix}. \tag{15.7}$$

Proof: A direct calculation shows that $\Psi \circ S = \widehat{S} \circ \Psi$ for the map $\widehat{S}$ given above. Once we have the map $\widehat{S}$, the given domain is clearly an invariant domain. □

15.5 COMPACTIFYING QUARTER TURN MAPS

In this section we do for quarter turn maps what we did for shears in the last section. We will deal with the map R_{0,r_n} in detail, then at the end explain the minor changes needed for the other cases.

The domain for our PET is $\mathbf{T}^n \times [-1/2, 1/2]$ but our PET is not defined on the entire domain. To get a domain on which the PET is defined, we take

$$X_{0,n} = \mathbf{T}^{n-1} \times (1/2, 1/2) \times (-1/2, 1/2). \tag{15.8}$$

When we use this domain, the formula for Ψ takes a particular form. Let int(t) be the integer nearest t and let $I(t) = t - \text{int}(t)$. The point $I(t)$ is a representative for t in the fundamental domain $[-1/2, 1/2]$ for the action of $\mathbf{Z}$ on $\mathbf{R}$. We will not worry about the boundary cases when t is a half-integer; they do not arise for points where our maps are well defined. We have

$$\Psi(x, y) = \left(\left[\frac{x}{r_1}, \cdots, \frac{x}{r_{n-1}} \right], I\left(\frac{x}{r_n} \right), y \right). \tag{15.9}$$

Now we are ready to state the main result.

Lemma 15.7 *Let* $R_n = R_{0,r_n}$. *There exists an affine PET* $\widehat{R}_n : \widehat{\mathbf{S}} \to \widehat{\mathbf{S}}$ *such that* $\Psi \circ R_n = \widehat{R}_n \circ \Psi$. *The linear part of* $\widehat{R}_n$ *is given by the matrix*

$$\begin{bmatrix} 1 & 0 & 0 & \cdots & -r_n/r_1 & r_n/r_1 \\ 0 & 1 & 0 & \cdots & -r_n/r_2 & r_n/r_2 \\ 0 & 0 & 1 & \cdots & -r_n/r_3 & r_n/r_3 \\ \cdots & & & & & \\ 0 & 0 & 0 & \cdots & 0 & 1 \\ 0 & 0 & 0 & \cdots & -1 & 0 \end{bmatrix}. \tag{15.10}$$

$\widehat{R}_n$ *is well defined on every point of* $X_{0,n}$ *and preserves* $X_{0,n}$.

Proof: We have a tiling of S by $r_n \times 1$ rectangles, one of which is centered at the origin. The map R_n affinely turns each rectangle a quarter turn clockwise. Let $R_{0,1}$ be the quarter turn map with parameters $(q, r) = (0, 1)$. This map is based on the tiling of S by unit squares centered at integer points, and $R_{0,1}$ just isometrically turns each square a quarter turn clockwise. In particular, if $(x_2, y_2) = R_{0,1}(x_1, y_1)$ then x_1 and x_2 have the same nearest integer.

We define

$$g(x, y) = (r_n x, y) \tag{15.11}$$

and introduce the point

$$x^* = r_n \ \mathrm{int}\left(\frac{x}{r_n}\right). \tag{15.12}$$

Given that $R_n = g \circ R_{0,1} \circ g^{-1}$, we have the following fact:

$$(x_2, y_2) = R_n(x_1, y_1) \implies x_2^* = x_1^*. \tag{15.13}$$

We introduce $F : X_{0,n} \to X_{0,n}$ and $\Psi^* : \mathbf{S} \to X_{0,n}$ as follows:

$$F = \mathrm{Id}_{n-1} \times R_{0,1}, \qquad \Psi^*(x, y) = \left(\left[\frac{x^*}{r_1}, \cdots, \frac{x^*}{r_{n-1}}\right], I\left(\frac{x}{r_n}\right), y\right). \tag{15.14}$$

First Calculation: A calculation shows that

$$\Psi^* \circ R_n = F \circ \Psi^*. \tag{15.15}$$

Here is an explanation. For the first $n - 1$ coordinates this formula just comes from the implication in Equation 15.13. Let π be projection on the last 2 coordinates. Letting $J(x, y) = (I(x), y)$, we have

$$\pi \circ \Psi^* \circ g = J, \qquad J \circ R_{0,1} \circ g^{-1} = \pi \circ F \circ \Psi^*.$$

Hence

$$\pi \circ \Psi^* \circ R_n = (\pi \circ \Psi^* \circ g) \circ R_{0,1} \circ g^{-1} = J \circ R_{0,1} \circ g^{-1} = \pi \circ F \circ \Psi^*.$$

So, the final 2 coordinates work out as well. This establishes our claim.

Second Calculation: We introduce the invertible map $G : X_{0,n} \to X_{0,n}$ as follows:

$$G([x_1, \cdots, x_{n-1}], x_n, y) = \left(\left[x_1 + \frac{r_n}{r_1} x_n, \cdots, x_{n-1} + \frac{r_n}{r_{n-1}} x_n \right], x_n, y \right).$$

$$(15.16)$$

Since Ψ and Ψ^* agree in the last two coordinates and G is the identity in the last two coordinates, we have $\Psi = G \circ \Psi^*$ in the last two coordinates. We will check this equality in the first coordinate. In the remaining coordinates the check is the same. The first coordinate of $G \circ \Psi^*(x, y)$ is the representative of t in $\mathbf{R}/\mathbf{Z}$, where

$$t = \frac{r_n \operatorname{int}(x/r_n)}{r_1} + \frac{r_1}{r_n} I(x/r_n) = \frac{r_1}{r_n} \left(\operatorname{int}(x/r_n) + I(x/r_n) \right) = \frac{r_1}{r_n} \times \frac{x}{r_n} = \frac{x}{r_1}.$$

But this is also the first coordinate of $\Psi(x, y)$. Hence

$$\Psi = G \circ \Psi^*, \qquad G^{-1} \circ \Psi = \Psi^*. \qquad (15.17)$$

The first equation implies the second.

Third Calculation: The domain $X_{0,n}$ is invariant under both G and F. So, consider the following map on $X_{0,n}$:

$$\widehat{R}_n = G \circ F \circ G^{-1}. \qquad (15.18)$$

We compute

$$\widehat{R}_n \circ \Psi = G \circ F \circ G^{-1} \circ \Psi = G \circ F \circ \Psi^* = G \circ \Psi^* \circ R_n = \Psi \circ R_n.$$

Thus $\widehat{R}_n$ is the map we seek. We have the formula

$$\widehat{R}_n([x_1, ..., x_{n-1}], x_n, y) =$$

$$\left(\left[x_1 - \frac{r_n}{r_1} x_n + \frac{r_n}{r_1} y, ..., x_{n-1} - \frac{r_n}{r_{n-1}} x_n + \frac{r_n}{r_{n-1}} y \right], y, -x_n \right). \qquad (15.19)$$

This leads to the form of the linear part of $\widehat{R}_n$ given in Equation 15.10. □

Now we deal with the other cases. For the map $R_{1,n}$ the natural domain to consider is

$$X_{1,n} = \mathbf{T}^{n-1} \times (0, 1) \times (-1/2, 1/2). \qquad (15.20)$$

The map $R_{1,n}$ is conjugate to $R_{0,n}$ by the translation $(x, y) \to (x + r_n/2, y)$. The corresponding map $\widehat{R}_{1,n}$ is conjugate to $R_{0,n}$ by a translation which moves the nth coordinate by $1/2$. So, we get the same result but with $X_{1,n}$ replacing $X_{0,n}$.

The same results hold for R_{q_j, n_j}. The only difference is that the roles played by the indices j and n are swapped. For instance, the linear part of $\widehat{R}_1$ is given by the matrix

$$\begin{bmatrix} 0 & 0 & 0 & \cdots & 0 & 1 \\ -r_1/r_2 & 1 & 0 & \cdots & 0 & r_1/r_2 \\ -r_1/r_3 & 0 & 1 & \cdots & 0 & r_1/r_3 \\ & & \cdots & & & \\ -r_1/r_n & 0 & 0 & \cdots & 1 & r_1/r_n \\ -1 & 0 & 0 & \cdots & 0 & 0 \end{bmatrix}, \qquad (15.21)$$

and the invariant domain is the image of $X_{q_1,n}$ under the map which swaps the 1st and nth coordinates.

15.6 THE END OF THE PROOF

Now we assemble the ingredients from the previous sections and prove Theorem 15.1. Referring to the maps in the previous two sections, we define

$$\widehat{\mathcal{T}} = \widehat{S}_{s_n} \circ \widehat{R}_{q_n,r_n} \circ \cdots \circ \widehat{S}_{s_1} \circ \widehat{R}_{q_1,r_1}. \tag{15.22}$$

The composition of affine pets is an affine PET. Hence $\widehat{\mathcal{T}}$ is an affine PET. By construction, $\Psi \circ \mathcal{T} = \widehat{\mathcal{T}} \circ \Psi$. The results in §15.3 now imply Statements 1 and 2 of Theorem 15.1.

Now we turn to Statement 3. Let $\mathbf{S}^*$ denote the closure of $\Psi(\mathbf{S})$ in $\widehat{\mathbf{S}}$. Suppose $\mathcal{T}^k$ is finitary for some exponent k. We will prove that the restriction of $\widehat{\mathcal{T}}^k$ to $\mathbf{S}^*$ is an ordinary PET. For ease of notation, we assume that $k = 1$. The proof works the same way regardless of exponent.

We need to show that $\widehat{\mathcal{T}}$ is locally a translation. Suppose $\widehat{p} \in \mathbf{S}^*$ and $\{\widehat{p}_n\}$ is a sequence in $\mathbf{S}^*$ converging to $\widehat{p}$. We will show that

$$\widehat{\mathcal{T}}(\widehat{p}) - \widehat{p} = \widehat{\mathcal{T}}(\widehat{p}_n) - \widehat{p}_n, \tag{15.23}$$

for all n sufficiently large. Since $\Psi(\mathbf{S})$ is dense in $\mathbf{S}^*$ and the linear part of $\widehat{\mathcal{T}}$ is independent of point, it suffices to consider the case when $\widehat{p} = \Psi(p)$ and $\widehat{p}_n = \Psi(p_n)$ for some $p \in \mathbf{S}$ and some sequence $\{p_n\}$ in $\mathbf{S}$. Note that $\{p_n\}$ need not be a convergent sequence in $\mathbf{S}$.

Lemma 15.8 *Setting $V_s = \mathcal{T}(s) - s$ for any $s \in S$, we have*

$$\widehat{\mathcal{T}}(\widehat{p}) - \widehat{p} = \Psi(V_p), \qquad \widehat{\mathcal{T}}(\widehat{p}_n) - \widehat{p}_n = \Psi(V_{p_n}). \tag{15.24}$$

Proof: We have

$$\widehat{\mathcal{T}}(\widehat{p}) - \widehat{p} = \widehat{\mathcal{T}} \circ \Psi(p) - \Psi(p) = \Psi \circ \mathcal{T}(p) - \Psi(p) = \Psi(V_p). \tag{15.25}$$

The last equality comes from the fact that $\Psi(V + W) = \Psi(V) + \Psi(W)$ whenever V, W, and $V + W$ all belong to S. Here we are taking $V = V_p$ and $W = W_p$. The same argument works for p_n. $\square$

We now observe the following properties.

1. We have $\Psi(V_{p_n}) = \widehat{\mathcal{T}}(p_n) - \widehat{p}_n$. Since $\widehat{p}_n \to \widehat{p}$ and Ψ is continuous in a neighborhood of $\widehat{p}$, we have $\widehat{\mathcal{T}}(\widehat{p}_n) \to \widehat{\mathcal{T}}(\widehat{p})$. Hence $\Psi(V_{p_n}) \to \Psi(V_p)$.

2. Since $\mathcal{T}$ is finitary, there are only finitely many choices for V_{p_n}.

These properties imply that $\Psi(V_{p_n}) = \Psi(V_p)$ for n large. This fact combines with Equation 15.24 to establish Equation 15.23 for n large. This completes the proof.

Chapter Sixteen

The Nature of the Compactification

16.1 CHAPTER OVERVIEW

In this chapter we will look more closely at Theorem 15.1 and give more information about the PET that appears in that result.

Sometimes an affine PET can be defined in terms of a triple (X_1, X_2, I), where X_1 and X_2 are polyhedral fundamental domains for $\mathbf{Z}^{n+1}$ and I is a linear isomorphism from X_1 to X_2. The affine PET is given by

$$[X_1, X_2, I] := \Pi_2 \circ I \circ \Pi_1^{-1}. \tag{16.1}$$

Here Π_j is the canonical map from X_j to $\widehat{\mathbf{S}}$. The map $\Pi_j^{-1} : \widehat{\mathbf{S}} \to X_j$ is defined as follows: Lift to $\mathbf{R}^{n+1}$ then translate by the appropriate integer vector. The first main goal of this chapter is to prove the following result.

Theorem 16.1 *The affine PET from Theorem 15.1 is conjugate to a map* $[X_1, X_2, I]$, *where* X_1 *and* X_2 *are parallelotope fundamental domains for* $\mathbf{Z}^{n+1}$, *centered at the origin. In case the QTC in Theorem 15.1 comes from an outer billiards system, the map* $[X_1, X_2, I]^2$ *is an ordinary PET and* I^2 *is an involution.*

We use the notation from Theorem 15.1. The basic idea of the proof is to remove from the torus $\widehat{\mathbf{S}}$ the *singular set* – i.e., the places where the PET is not defined. We will see that what is left over is isometric to the interior of a convex parallelotope. We analyze the singular set in §16.2. In §16.3 we construct X_1. In §16.4 we construct the second parallelotope based on the action of the PET from Theorem 15.1. Our proof of Theorem 16.1 finishes at the end of §16.4.

In proving Theorem 16.1 we will keep the notation from Theorem 15.1. In particular, $\mathcal{T}$ is the quarter turn composition and $\widehat{\mathcal{T}}$ is the PET. The maps $\widehat{R}_j$ and $\widehat{S}_j$ are the compactifications of the maps R_j and S_j comprising $\mathcal{T}$. These maps are defined in Lemmas 15.6 and 15.7 respectively.

In §16.5 we will restate the case of Theorems 15.1 and 16.1 that apply to the pinwheel map associated to outer billiards on a polygon without parallel sides. Our result is Theorem 16.9. Finally, in §16.7 we see how Theorems 0.4 and 16.9 match up.

16.2 THE SINGULAR DIRECTIONS

Let $\mathcal{H}$ denote a finite union of n-dimensional linear subspaces of $\mathbf{R}^{n+1}$. We say that $\mathcal{H}$ is a *complete set* for the map $\widehat{\mathcal{T}}$ if $\widehat{\mathcal{T}}$ is defined on $\widehat{\mathbf{S}} - X$, where X is a finite union of codimension 1 flat tori and each element of X is parallel to some element of $\mathcal{H}$. In this section we will produce a complete set $\mathcal{H}$ with $n+1$ members. This is a first step towards proving Theorem 16.1 because the map $[X_1, X_2, I]$ discussed in Theorem 16.1 has a complete set with $n+1$ members.

Let Π_k denote the hyperplane given by the equation $x_k = 0$. To keep our notation consistent with the previous chapter, we say that Π_{n+1} is the hyperplane given by $y = 0$. Let dF denote the linear part of the affine map F. Note that $d(F^-) = (dF)^{-1}$ so there is no ambiguity in writing dF^{-1}.

Lemma 16.2 *A complete set for $\widehat{\mathcal{T}}$ is given by*

1. $\{\Pi_1, \Pi_{n+1}\}$,

2. $d\widehat{R}_1^{-1}(\Pi_{n+1})$,

3. $d\widehat{R}_1^{-1}d\widehat{S}_1^{-1}(\{\Pi_2, \Pi_{n+1}\})$,

4. $d\widehat{R}_1^{-1}d\widehat{S}_1^{-1}d\widehat{R}_2^{-1}(\Pi_{n+1})$,

5. $d\widehat{R}_1^{-1}d\widehat{S}_1^{-1}d\widehat{R}_2^{-1}d\widehat{S}_2^{-1}(\{\Pi_3, \Pi_{n+1}\})$,

and so on.

Proof: In view of Lemma 15.6, the hyperplane Π_{n+1} is a complete set for $\widehat{S}_j$. In view of Lemma 15.7 (and the discussion after it), the two hyperplanes $\{\Pi_k, \Pi_{n+1}\}$ form a complete set for $\widehat{R}_k$. If the map $\widehat{\mathcal{T}}$ is not defined on some point p, then one of the compositions

$$F_k = \widehat{R}_k \circ \widehat{S}_{k-1} \circ \cdots \circ \widehat{S}_1 \circ \widehat{R}_1, \qquad G_k = \widehat{S}_k \circ \widehat{R}_k \circ \cdots \circ \widehat{S}_1 \circ \widehat{R}_1 \quad (16.2)$$

is undefined at p but all shorter compositions are defined. But then either $F_k(p)$ lies in the boundary of the invariant domain for $\widehat{R}_k$ or $G_k(p)$ lies in the boundary of the invariant domain for $\widehat{S}_k$. But then p lies in a hypersurface parallel to one of the hyperplanes on our list. $\square$

Note that there are about $3n$ hyperplanes listed in Lemma 16.2 whereas we are claiming that $n+1$ hyperplanes suffices. The idea in the next lemma is just to eliminate redundancies.

Lemma 16.3 *A complete list for $\widehat{\mathcal{T}}$ is given by $H_1, ..., H_{n+1}$, where $H_1 = \Pi_1$ and $H_{n+1} = \Pi_{n+1}$ and*

$$H_{k+1} = d\widehat{R}_1^{-1} \circ ... \circ d\widehat{S}_k^{-1}(\Pi_{k+1}), \qquad k = 1, ..., n-1. \quad (16.3)$$

Proof: First, we have

$$d\widehat{R}_k^{-1}(\Pi_{n+1}) = \Pi_k. \tag{16.4}$$

Therefore, each hyperplane listed on line $2k$ of Lemma 16.2 is contained in one of the hyperplanes listed on line $2k - 1$.

Second, we have

$$dS_k^{-1}(\Pi_{n+1}) = \Pi_{n+1}, \qquad d\widehat{R}_k^{-1}(\Pi_k) = \Pi_{n+1}. \tag{16.5}$$

Therefore, the second hyperplane listed on line $2k + 1$ of Lemma 16.2 is contained in the first hyperplane listed on line $2k - 1$. For instance, taking $k = 2$, we have

$$d\widehat{R}_1^{-1}d\widehat{S}_1^{-1}d\widehat{R}_2^{-1}d\widehat{S}_2^{-1}(\Pi_{n+1}) =$$

$$d\widehat{R}_1^{-1}d\widehat{S}_1^{-1}d\widehat{R}_2^{-1}(\Pi_{n+1}) = d\widehat{R}_1^{-1}d\widehat{S}_1^{-1}(\Pi_2).$$

Upon eliminating all the redundancies, we get the advertised list. $\square$

Let e_k denote the kth standard basis vector in $\mathbf{R}^{n+1}$. Let $H_k^{\perp}$ denote the normal to H_k.

Lemma 16.4 *The matrix whose rows are $H_1^{\perp}, ..., H_{n+1}^{\perp}$ has determinant 1.*

Proof: Let $M_k = d\widehat{R}_1 \circ ... \circ d\widehat{S}_{k-1}$. We have $H_{n+1}^{\perp} = (0, ..., 0, 1)$ and

$$H_k^{\perp} = (M_k^{-1})^t(e_k), \qquad k = 1, ..., n. \tag{16.6}$$

The maps $d\widehat{R}_j$ and $d\widehat{R}_k$ act trivially on $e_{j+1}, ..., e_n$. Hence M_k acts trivially on $e_{k+1}, ..., e_n$. Hence, rows $k, ..., n$ of the inverse transpose matrix $(M_k^{-1})^t$ coincide with the rows of the identity matrix. Hence

$$H_k^{\perp} = (*, \cdots, *, 1, 0, \cdots, 0, *), \qquad k = 1, ..., n. \tag{16.7}$$

The 1 appears in the kth slot and $(*)$ indicates an entry that we don't explicitly know. The lemma is immediate from this structure. $\square$

16.3 THE FIRST PARALLELOTOPE

Let $X_1 \subset \mathbf{R}^{n+1}$ be the parallelotope consisting of vectors V such that

$$H_i^{\perp} \cdot V \in [-1/2, 1/2] \tag{16.8}$$

for all i. Given the form of the matrix in Equation 16.9 below, we can also say that X_1 is the parallelotope bounded by the hyperplanes $H_k \pm 1/2e_k$.

Lemma 16.5 *X_1 is a fundamental domain for $\mathbf{Z}^{n+1}$.*

Proof: In view of Lemma 16.4, the set X_1 is a unit volume parallelotope. Let M be the matrix with rows $H_1^\perp, ..., H_{n+1}^\perp$. From the proof of Lemma 16.4 we have

$$
M = \begin{bmatrix}
1 & 0 & 0 & \cdots & 0 & * \\
* & 1 & 0 & \cdots & 0 & * \\
* & * & 1 & \cdots & 0 & * \\
\cdots & & & & & \\
* & * & * & \cdots & 1 & * \\
0 & 0 & 0 & \cdots & 0 & 1
\end{bmatrix}.
\tag{16.9}
$$

X_1 consists of those vectors $V \in \boldsymbol{R}^{n+1}$ such that $M(V) \in [-1/2, 1/2]^{n+1}$.

Since X_1 has unit volume, it suffices to show that the interior of X_1 does not intersect some integer translate of X_1. This happens if and only if there is some integer vector $V \in \boldsymbol{Z}^{n+1}$ such that $MV \in (0,1)^{n+1}$. This is impossible, as we now explain. Suppose $M(x_1, ..., x_{n+1}) \in (0,1)^{n+1}$. The last coordinate is just x_{n+1}. Hence $x_{n+1} = 0$. But then the action of M on the first n coordinates coincides with the action of a lower triangular matrix with 1s along the diagonal. $\square$

Let $q = (q_1, ..., q_n)$ and $r = (\cdots)$ and $s = (\cdots)$ be the invariants for $\mathcal{T}$. Let $\pi : \boldsymbol{R}^{n+1} \to \widehat{\boldsymbol{S}}$ be projection. Let X_1^o be the interior of X_1. Let I be the affine map which fixes the vector $q/2$ and whose linear part coincides with the linear part of $\widehat{\mathcal{T}}$.

Lemma 16.6 *The map $\widehat{\mathcal{T}}$ is entirely defined on $\pi(X_1^o + q/2)$. Furthermore,*

$$
\widehat{\mathcal{T}} = \pi \circ I \circ \pi^{-1}
$$

on $\pi(X_1^o + q/2)$ provided that π^{-1} is taken to have its range in $X_1^o + q/2$.

Proof: We will give the proof in case $q = (0, ..., 0)$. In this case, I is simply the linear part of $\widehat{\mathcal{T}}$. The general case has essentially the same proof, and differs only in that we apply suitable translations to the basic objects.

Let A_k denote the open slab bounded by the hyperplanes $x_k = \pm 1/2$. Let B_k denote the open slab bounded by the parallel hyperplanes $H_k \pm \frac{1}{2}e_k$. By construction $X_1 = \bigcap H_k$. Also by construction,

$$
d\widehat{R}_1^{-1} \circ \cdots \circ d\widehat{S}_{k-1}^{-1}(A_k) = B_k.
\tag{16.10}
$$

Let ρ_k be the restriction of $d\widehat{R}_k$ to $A_k \cap A_{n+1}$. Likewise, let σ_k be the restriction of $d\widehat{S}_k$ to A_{n+1}. Given the description of the invariant domains for $\widehat{S}_k$ and $\widehat{R}_k$ in §15.4 and §15.5, we have

$$
\widehat{R}_k = \pi \circ \rho_k \circ \pi^{-1}, \qquad \widehat{S}_k = \pi \circ \sigma_k \circ \pi^{-1}.
\tag{16.11}
$$

The right-hand side is independent of the lift, as long as the range of π^{-1} is taken to be $A_k \cap A_{n+1}$ or A_{n+1} respectively.

Choose any point $p \in \pi(X_1^o)$. Let q_1 be the unique point in X_1^o such that $\pi(q_1) = p$. By construction $q_1 \in B_1 \cap ... \cap B_{n+1}$. But $A_1 = B_1$ and

$A_{n+1} = B_{n+1}$. Hence $q_1 \in A_1 \cap A_{n+1}$. Since $q_1 \in A_1 \cap A_{n+1}$, the map ρ_1 is defined on q_1. Since ρ_1 preserves $A_1 \cap A_{n+1}$, we have $\rho_1(q_1) \in A_1 \cap A_{n+1}$. In particular $\rho_1(q_1) \in A_{n+1}$, and so σ_1 is defined on $\rho_1(q_1)$. Equation 16.11 now gives us

$$q_2 = \sigma_1 \circ \rho_1(q_1) \in A_2 \cap A_{n+1}, \qquad \pi(q_2) = \widehat{S}_1 \circ \widehat{R}_1(p). \qquad (16.12)$$

Repeating the same argument with q_2 in place of q_1, we see that ρ_2 is defined on q_2 and σ_2 is defined on $\rho_2(q_2)$ and

$$q_3 = \sigma_2 \circ \rho_2(q_2) \in A_3 \cap A_{n+1}, \qquad \pi(q_3) = \widehat{S}_2 \cap \widehat{R}_2 \circ \widehat{S}_1 \circ \widehat{R}_1(p). \quad (16.13)$$

Continuing in this way, we produce points $q_4, ..., q_n$ such that

- $q_k \in A_k \cap A_{n+1}$.

- $\sigma_k \circ \rho_k$ is defined on q_k.

- $q_{k+1} = \sigma_k \circ \sigma_k(q_k)$.

- $\pi \circ q_{k+1} = \widehat{S}_k \circ \widehat{R}_k(\pi(q_k))$.

In particular, $\widehat{\mathcal{T}}$ is defined on p and

$$\widehat{\mathcal{T}}(p) = \pi(q_n) = \pi \circ \sigma_n \circ \cdots \circ \rho_1 \circ \pi^{-1}(q_1) = I \circ \pi^{-1}(p). \qquad (16.14)$$

Hence $\widehat{\mathcal{T}}$ is completely defined on $\pi(X_1^o)$ and $\widehat{\mathcal{T}} = \pi \circ I \circ \pi^{-1}$ on $\pi(X_1^o)$. $\square$

16.4 THE SECOND PARALLELOTOPE

Let $X_2 = I(X_1)$.

Lemma 16.7 X_2 is a fundamental domain for $\mathbf{Z}^{n+1}$.

Proof: Again, we consider the case when $q = (0, ..., 0)$ for ease of exposition. The linear parts of $\widehat{R}_k$ and $\widehat{S}_k$ are orientation preserving and volume preserving maps. We also know that X_1 is a unit volume parallelotope and a fundamental domain for $\mathbf{Z}^{n+1}$. Since I is volume preserving, X_2 is also a unit volume parallelotope.

The map $\widehat{\mathcal{T}}$ is invertible. In particular, the restriction of $\widehat{\mathcal{T}}$ to X_1^o is injective. But this map equals $\pi \circ I \circ \pi^{-1}$. Hence $\pi : X_2 \to \widehat{\mathbf{S}}$ is also injective. This fact, together with the fact that X_2 has unit volume, shows that X_2 is in fact a fundamental domain. $\square$

When $q = (0, ..., 0)$, Lemma 16.6 tells us that $\widehat{\mathcal{T}} = [X_1, X_2, I]$. In general, let $X_j' = X_j + q/2$ and let I' be the affine map which fixes $q/2$ and whose linear part is I. Lemma 16.6 tells us that $\widehat{\mathcal{T}} = \pi \circ I' \circ \pi^{-1}$ on the interior of $\pi(X_1')$. But $[X_1', X_2', I']$ is conjugate to $[X_1, X_2, I]$.

More precisely, let $\tau : \widehat{\mathbf{S}} \to \widehat{\mathbf{S}}$ be translation by $q/2$. Then

$$\tau \circ [X_1, X_2, I] \circ \tau^{-1} = [X_1', X_2', I'].$$

It is convenient to define the new map

$$\Psi_q = \tau^{-1} \circ \Psi = \Psi \pm q/2. \qquad (16.15)$$

(Since we are dividing out by $\mathbf{Z}^{n+1}$ it doesn't matter if we take $+q/2$ or $-q/2$. We get the same map.) A short calculation tells us that

$$\Psi_q \circ \mathcal{T} = [X_1, X_2, I] \circ \Psi_q. \qquad (16.16)$$

From this alternate point of view, the compactified system $[X_1, X_2, I]$ is independent of the q parameters. What changes with the q parameters is the map Ψ_q.

The next lemma finishes the proof of Theorem 16.1.

Lemma 16.8 *Suppose that* $\mathcal{T}$ *is a QTC that arises from outer billiards. Then* $[X_1, X_2, I]^2$ *is an ordinary PET and the map* I *is an involution.*

Proof: We set $F = [X_1, X_2, I]$. This is the PET from Theorem 16.1. According to Lemma 14.5, the map $\mathcal{T}^2$ is finitary. But this means that F^2 is a piecewise translation, an ordinary PET. The linear part of F^k is I^k for any k. In particular, this is true for $k = 2$. Since F^2 is an ordinary PET, the linear part of F^2 is the identity. In short, I^2 is the identity. That is, I is an involution. $\square$

Remark: With more work, we could show that I has $n - 1$ eigenvalues which are $+1$ and 2 eigenvalues which are -1.

16.5 THE GENERAL MASTER PICTURE THEOREM

Now we specialize Theorems 15.1 and 16.1 to the case of polygonal outer billiards. There is nothing new in the result below. It is just a reformulation of this special case. We will state our result below in terms of the pinwheel map because this is the part for which we have given a self-contained proof. Theorem 14.1 relates the outer billiards map to the pinwheel map and the easier Lemma 14.6 gives an even stronger result for kites.

Let P be a convex polygon without parallel sides. Let $\Sigma_1, ..., \Sigma_n$ be the strips associated to P as in §14.2. Let $\alpha_k = A_k/A_n$ where A_k is the area of $\Sigma_k \cap \Sigma_{k+1}$. Let

$$\text{rank}(P) = 1 + \dim \boldsymbol{Q}(\alpha_1, ..., \alpha_n). \qquad (16.17)$$

For the dimension count, we think of $\boldsymbol{Q}(\alpha_1, ..., \alpha_n)$ as a vector space over $\boldsymbol{Q}$. Theorems 15.1 and 16.1 immediately give the following result.

Theorem 16.9 (Graph Master Picture Theorem) *Let P be a convex n-gon without parallel sides. Let Σ be one of the strips associated to P. Let F be the pinwheel map relative to Σ. There is an affine PET $\widehat{F} = [X_1, X_2, I]$ acting on $\widehat{\Sigma} = T^{n+1}$ and a locally affine map $\Psi : \Sigma \to \widehat{\Sigma}$ such that $\widehat{F} \circ \Psi = \Psi \circ F$. The map $\widehat{F}^2$ is an ordinary PET. The closure of $\Psi(\Sigma)$ is a flat sub-torus of dimension* rank(P).

16.6 STRUCTURE OF THE PET

Let $\widehat{F} = [X_1, X_2, I]$. We consider the case when I is an involution and $\widehat{F}^2$ is an ordinary PET defined on T^{n+1}. We set $\Lambda = I(Z^n)$. Note that Λ is another lattice in R^{n+1}. Since X_1 is a fundamental domain for Z^{n+1} and $X_2 = I(X_1)$, we see that X_1 is a fundamental domain for Λ. Likewise, X_2 is a fundamental domain for Λ. This means that $\Lambda(X_1)$ and $\Lambda(X_2)$ are both tilings of R^{n+1} by parallelopipeds. These parallelopipeds need not meet face to face however.

Lemma 16.10 *The following is true.*

1. *The forward partition for $\widehat{F}^2$ is $\Pi_1(X_1 \cap \Lambda(X_2))$.*

2. *The backward partition for $\widehat{F}^2$ is $\Pi_2(X_2 \cap \Lambda(X_1))$.*

Proof: We prove Statement 1. Statement 2 has essentially the same proof. Let $\Pi : R^{n+1} \to T^{n+1}$ be the quotient map. We have $\Pi_j = \Pi|_{X_j}$ for $j = 1, 2$. We compute

$$\widehat{F}^2 = \Pi_2 \circ (I \circ \Pi_1^{-1} \circ \Pi_2 \circ I) \circ \Pi_1^{-1} =$$

$$\Pi_2 \circ (I \circ \phi_{X_2} \circ I) \circ \Pi_1^{-1} =$$

$$\Pi_2 \circ \phi|_{X_1} \circ \Pi_1^{-1} =$$

$$\Pi \circ \phi|_{X_1} \circ \Pi_1^{-1}. \tag{16.18}$$

Starting with $x \in T^{n+1}$, we let $\widehat{x} = \Pi_1^{-1}(x)$. We find $\widehat{x}$ by seeing how x sits with respect to the domain $\Pi_1(X_1)$ and then we pull back. Next, we add a suitable vector $V_x \in \Lambda_2$ so that $\widehat{x} + V_x \in X_2$. Then we project back to T^{n+1}. This gives

$$\widehat{F}^2(x) = \Pi(\widehat{x} + V_x) = x + \Pi(V_x).$$

Therefore, $\widehat{F}^2(x)$ only depends on V_x, which is determined by the location of $\widehat{x}$ in the tiling $\Lambda_2(X_2)$. So, the partition for $\widehat{F}^2$ is $\Pi_1(X_1 \cap \Lambda_2(X_2))$. $\square$

Slicing the Picture: Let Ω be the partition of T^{n+1} corresponding to the forward direction of $\widehat{F}^2$. Here we discuss how we understand 2-dimensional slices of Ω. Let $\Upsilon \subset R^{n+1}$ be some 2-plane. We are interested in how the

$\Pi(\Upsilon)$ and $\Pi(\Omega)$ intersect. When we lift this intersection to the universal cover $\boldsymbol{R}^{n+1}$, we get

$$\Upsilon \cap \bigcup_{V \in \boldsymbol{Z}^{n+1}} \Omega + V. \qquad (16.19)$$

In the case of interest, $\Pi(\Upsilon)$ is a 2-torus, and the polygonal tiling in Equation 16.19 is periodic. It turns out that the simpler union

$$\Upsilon \cap \bigcup_{i \in \{-1,0,1\}} \Omega + (0,0,0,i,0) \qquad (16.20)$$

always comprises a single square fundamental domain for the tiling.

Generalized Arithmetic Graph: There is a general kind of arithmetic graph associated to $\widehat{F}^2$. Referring to the proof of Lemma 16.10, the vector V_x belongs to $I(\boldsymbol{Z}^{n+1})$, so $I(V_x) \in \boldsymbol{Z}^{n+1}$. Given an orbit $x = x_0, x_1, x_2, \dots$ under $\widehat{F}^2$ we associate the path $\{\Gamma_k(x)\}$ where

$$\Gamma_k(x) = \sum_{j=0}^{k-1} I(V_{x_j}). \qquad (16.21)$$

We call this path $\Gamma(x)$. We can get a more global view by choosing an offset vector and proceeding as in the definition of the ordinary arithmetic graph given in §13.3, but we leave this to the interested reader.

16.7 THE CASE OF KITES

Here we explain why the case of Theorem 16.9 which applies to special orbits on kites is essentially equivalent to Theorem 0.4. We fix a parameter $A = p/q$. The map

$$f(m,n) = (2Am + 2n + 1/q, (-1)^{m+n+1}) \qquad (16.22)$$

conjugates the curve-following dynamics on the arithmetic graph Γ to the map Θ acting on the centers of the special intervals in $\boldsymbol{R} \times \{-1,1\}$. Pulling back by f, we see that Theorem 0.4 is equivalent to the statement that some locally affine map intertwines the action of Θ with the action of a certain 3-dimensional PET. As we discussed in §14.6, the map Θ is conjugate to $\mathcal{T}$, the QTC. Thus, we can further reinterpret Theorem 0.4 as saying that there is a locally affine map which intertwines $\mathcal{T}$ with the action of a 3-dimensional PET. Theorem 16.9 also says this.

Now we describe what Theorem 16.9 gives us in the special case of kites, and then we line this up with what we get from Theorem 0.4.

The Map: The map $\mathcal{T}$ is as in §14.6. Given Equation 14.18, the map $\Psi_q : \mathbf{S} \to \widehat{\mathbf{S}}$ is given by

$$\Psi_q(x,y) = \left[\left(x, \frac{1-A}{1+A}x, \frac{4A}{(1+A)^2}x, \frac{1-A}{1+A}x, y\right) + \left(\frac{1}{2}, \frac{1}{2}, 0, \frac{1}{2}, 0\right)\right]. \qquad (16.23)$$

Letting $\Delta \subset \mathbf{S}$ be as in Figure 14.7, we see that $\Psi_q(\Delta)$ is contained in the flat sub-torus of $U \subset \mathbf{T}^5$ given by the equations $x_2 = x_4$ and $2x_1 - 2y = 0$. For any A which is neither rational nor quadratic irrational, $\Psi_q(\Delta)$ is dense U.

The PET: Matrix calculations using the formulas in Lemma 15.4 and Lemma 15.5 show that the involution I is given by

$$I = \begin{bmatrix} 0 & -1 & \frac{-1-A}{2} & 0 & 0 \\ \frac{A-1}{A+1} & \frac{2A}{1+A} & \frac{A-1}{2} & 0 & 0 \\ \frac{-4A}{(1+A)^2} & \frac{-4A}{(1+A)^2} & \frac{1-A}{1+A} & 0 & 0 \\ \frac{A-1}{A+1} & \frac{A-1}{A+1} & \frac{A-1}{2} & 1 & 0 \\ 1 & 0 & \frac{-1-A}{2} & -1 & -1 \end{bmatrix}. \qquad (16.24)$$

Similar calculations show that X_j consists of those vectors v such that $M_j(v) \in [-1/2, 1/2]^5$. Here M_1 and M_2 respectively are

$$\begin{bmatrix} 1 & 0 & 0 & 0 & 0 \\ \frac{2A}{1+A} & 1 & 0 & 0 & \frac{1-A}{1+A} \\ \frac{-2A}{1+A} & \frac{1-A}{1+A} & 1 & 0 & 1 \\ -1 & 0 & \frac{1+A}{2} & 1 & 1 \\ 0 & 0 & 0 & 0 & 1 \end{bmatrix} \qquad \begin{bmatrix} 0 & -1 & \frac{-A-1}{2} & 0 & 0 \\ 0 & 0 & -1 & \frac{A-1}{A+1} & \frac{A-1}{A+1} \\ 0 & 0 & 0 & -1 & -1 \\ 0 & 0 & 0 & 0 & -1 \\ 1 & 0 & \frac{-A-1}{2} & -1 & -1 \end{bmatrix}. \qquad (16.25)$$

These matrices are such that $M_j I = M_{j+1}$, with indices taken mod 2.

The Arithmetic Graph: Consider the linear projection $H : \mathbf{Z}^5 \to \mathbf{Z}^2$ given by

$$H(x_1, x_2, x_3, x_4, x_5) = (x_3, x_2 + x_3 + x_4). \qquad (16.26)$$

It turns out that the component of the arithmetic graph associated to a point $p \in \Delta$ is obtained as follows. Let $x = \Psi_A(p) \in \mathbf{T}^5$. Let $\Gamma(x)$ be the generalized arithmetic graph for x. Then $H(\Gamma(x))$ is a translate of the ordinary arithmetic graph associated to the point $p' \in \mathbf{R} \times \{-1, 1\}$ corresponding to p under the conjugacy discussed in §14.6.

Matching the Partitions: We first discuss the picture for Theorem 16.9. Let $p \in \Delta$ be a point. Let $\Upsilon \subset \mathbf{R}^5$ be the 2-plane through $\Pi_1^{-1}(x)$ and spanned by

$$V_1 = \left(0, 1, \frac{A-1}{A+1}, 1, 0\right), \qquad V_2 = \left(0, 0, 1, 0, 0\right). \qquad (16.27)$$

Here $x = \Psi_A(p)$. Let $L : \mathbf{R}^5 \to \mathbf{R}^2$ be the linear transformation such that $L(V_1) = (1, 0)$ and $L(V_2) = (0, 1)$. Let τ be the image of the intersection from Equation 16.20 under the map L. Again, τ is a polygon-tiled square that serves as a fundamental domain for a periodic tiling.

Now we discuss the picture for Theorem 0.4. Let $p' \in \mathbf{R} \times \{-1, 1\}$ be the point corresponding to p under the conjugacy discussed in §14.6. Let $\Upsilon'_{p'}$

denote the YZ plane in $\boldsymbol{R}^3$ through the point $\Pi_1^{-1}(x')$ where $x' = \Psi'_A(p')$. Here Π'_1 is the map from the slice Y_A to the fundamental domain

$$\Omega' = [-1, 1] \times [1 + A]^2.$$

Let τ' be the intersection $\Upsilon' \cap \Omega'$ translated parallel to itself so that it lies in $\boldsymbol{R}^2$. Again, τ' is a square fundamental domain for a planar tiling.

The two pictures τ and τ' are always the same up to scaling. The two polygonal tilings of $\boldsymbol{R}^2$ are also identical up to scaling. Finally, the dynamical labels on the polygons are the same. In other words, if $\Psi_A(p)$ lands in a polytope which assigns the vector $\zeta \in \boldsymbol{Z}^5$ in the generalized arithmetic graph, then $\Psi'_A(p')$ lands in a polyhedron which assigns the vector $\zeta' = H(\zeta)$ in the ordinary arithmetic graph.

Computer Tie-In: Follow the instructions for the computer tie-in from §15.1. On the *PET slicer* window, press the *X slice match* button and set the *tile extent* option to *medium*. Set the partition to *forward*. The *PET* window then displays τ. If you use the magenta slider in the *PET slicer* window you can translate the slicing plane parallel to itself so as to explore all of the 3-dimensional partition $\Omega \cap U$. Set the *tile extent* to *large* if you want to see more of the tiling from Equation 16.19.

At the same time, open up the main program and look at the *graph PET* window. The *graph PET* window then shows τ'. Make sure that the *parameter* option is set to *locked* and the *slice* is set to X. Also, make sure that the parameter in the main program is the same as the one in the auxiliary program. Both are initially set to 2/9, but if you want to change parameters on the main program, you have to implement it in the *graph PET slicer* by pushing the *grab parameter* button.

In the auxiliary program, the magenta slider on the bottom of the *PET slicer* window slices through the PET from Theorem 16.9 in the same way that the *graph PET slicer* on the main program slices through the PET from Theorem 0.4 at the corresponding parameter. When the position of the magenta slider lies to the left of center, you need to use the $(+)$ partition in the main program. When the position of the magenta slider lies to the right, you need to use the $(-)$ partition. This is in accord with the construction in §13.5. When the sliders are in corresponding positions on the two programs, the pictures of τ and τ' match.

You can also see the component of the arithmetic graph plotted by the method described above. You do this by opening the *graph* window on the auxiliary program. Whenever you plot a tile in the QTC window, the corresponding component of the arithmetic graph is plotted in the graph window. It is translated in such a way that the origin corresponds to the selected point in the QTC window. You can see that these components match the ones shown in the main program, up to an affine transformation.

Part 4. The Plaid-Graph Correspondence

Chapter Seventeen

The Orbit Equivalence Theorem

17.1 CHAPTER OVERVIEW

This chapter begins Part 4 of the monograph. The goal of this part is to prove the Orbit Equivalence Theorem and the Quasi-Isomorphism Theorem. For convenience, we restate the Orbit Equivalence Theorem here. Let $F_X : X \to X$ and $F_Y : Y \to Y$ be the plaid PET and the graph PET respectively.

Theorem 17.1 (Orbit Equivalence) *There is a dynamically large subset $Z \subset X$ and a map $\Omega : Z \to Y$ with the following property. For any $\zeta \in Z$ with a well-defined orbit the following three statements hold:*

1. *There is some $k = k(\zeta) \in \{1, 2\}$ such that*

$$\Omega \circ F_X^k(\zeta) = F_Y \circ \Omega(\zeta). \tag{17.1}$$

2. *If $F_X(\zeta) \in Z$ then there is some $\ell = \ell(\zeta) \in \{0, 1\}$ such that*

$$\Omega \circ F_X(\zeta) = F_Y^\ell \circ \Omega(\zeta). \tag{17.2}$$

3. *ζ is a fixed point of F_X if and only if $\Omega(\zeta)$ is a fixed point of F_Y.*

The set Z is a union of 2 open convex integral prism quotients and the restriction of Ω to each one is an integral projective transformation that maps the slice Z_P into the slice Y_A, where $P = 2A/(1 + A)$. Finally, Ω is at most 2-to-1 and $\Omega(Z)$ is open dense in Y and contains all the well-defined F_Y-orbits.

In §17.2 we define Z. In §17.3 we define Ω. In §17.4 we characterize the image $\Omega(Z)$. In §17.5 we define a partition of Z into small convex polytopes which have the property that all the maps in Equations 17.1 and 1 are entirely defined and projective on each polytope. This allows us to verify the properties in the Orbit Equivalence Theorem just by checking what the two relevant maps do to the vertices of the new partition.

In §17.6 we put everything together and prove the Orbit Equivalence Theorem modulo some integer computer calculations. In §17.7 we discuss the computational techniques we use to carry out the calculations from §17.6. In §17.8 we explain the calculations.

17.2 THE PRISMS

As in the introduction, a *prism* is a set of the form $H \times \boldsymbol{R} \times [0,1]$ where H is a convex polygon. A *prism quotient* is the quotient of a prism by the cyclic group generated by the translation $(x, y, z, P) \to (x, y, z + 2, P)$.

We set

$$\widehat{X} = \widehat{Y} = \boldsymbol{R}^3 \times [0,1]. \tag{17.3}$$

These spaces are the universal covers of X and Y respectively. Even though they are the same space, it is useful to keep their names separate. In this section we just work with $\widehat{X}$.

Let Λ_X be the group of affine transformations such that $X = \widehat{X}/\Lambda_X$, as in §8.2. This is the group that is generated by the maps

- $T_1^2(x, y, z, P) = (x + 4, y + 2P, y + 2P, P)$

- $T_2(x, y, z, P) = (x, y + 2, z, P)$

- $T_3(x, y, z, P) = (x, y, z + 2, P)$.

Let $\pi_Z : \boldsymbol{R}^4 \to \boldsymbol{R}^3$ be the projection which drops the Z-coordinate:

$$\pi_Z(x, y, z, P) = (x, y, P). \tag{17.4}$$

The set Z is the union $Z_+ \cup Z_-$ of two convex integral prism quotients whose lifts $\widehat{Z}_+$ and $\widehat{Z}_-$ are both T_3-invariant. To describe Z_+ it suffices to list the vertices of $\pi_Z(\widehat{Z}_+)$:

$$(\pm 1, -1, 0), \ (2 \pm 1, 1, 0), \ \left(\frac{-1}{2} \pm \frac{1}{2}, -1, 1\right), \ \left(\frac{5}{2} \pm \frac{1}{2}, 2, 1\right). \tag{17.5}$$

Slicing this polyhedron with a plane $P = $ constant, we get a parallelogram. The shape of the parallelogram varies with the constant but its sides always have slope 0 and 1 and the left edge always lies in the line $X = Y$. See Figure 17.1 below. We define

$$Z_- = I(Z_+), \qquad I(x, y, z, P) = (-x, -y, -z, P). \tag{17.6}$$

The orbit $\Lambda_X(\widehat{Z})$ is an infinite union of convex integral prisms having disjoint interiors. This union gives a partial tiling of $\boldsymbol{R}^3 \times [0,1]$. The complementary pieces are also convex prisms.

Figure 17.1 below shows a picture of the Z slice at the parameter $P = 1/2$. The light and dark shaded domains are slices of the Λ_X-orbits of $\widehat{Z}_+$ and $\widehat{Z}_-$. The two unshaded regions labeled $+$ and $-$ are slices of the two components $\widehat{Z}'_{\pm}$ of $\widehat{X} - \widehat{Z}$ which intersect the fundamental domain

$$[-2, 2] \times [-1, 1]^2 \times [0, 1].$$

This fundamental domain is indicated in Figure 17.1 by a rectangle with arrows on its boundary.

The 5 vertices of $\pi_Z(\widehat{Z}'_+)$ are

$$(1, -1, 0), \ \left(\frac{3}{2} \pm \frac{1}{2}, 0, 1\right), \ \left(\frac{-1}{2} \pm \frac{1}{2}, -1, 1\right).$$

Finally $\widehat{Z}'_- = I(\widehat{Z}'_+)$. Let $Z'_{\pm}$ denote the projection of $\widehat{Z}'_{\pm}$ into X.

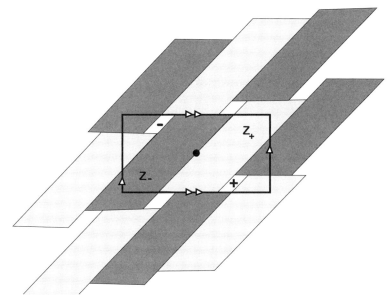

Figure 17.1: The Z slice at $P = 1/2$.

The arrows in Figure 17.1 indicate how the elements $T_1^2, T_2 \in \Lambda_X$ glue up our fundamental domain in a way that respects the tiling of $\widehat{X}$ by the Λ_X orbits of our 4 prisms. Hence, the quotient space X is partitioned into the convex prism quotients $Z_+ \cup Z_- \cup Z'_+ \cup Z'_-$.

Computer Tie-In: On the main program, open up the *plaid PET* window and the *plaid PET* control panel. On the *plaid PET slicer* set the *slice* to Z. On the *plaid PET* control panel you can turn on the displays of the sets considered above. By dragging the middle mouse button over the *plaid PET slicer* you can see pictures like Figure 17.1 for any parameter.

17.3 THE MAP

We first define a map

$$\widehat{\Omega} : \widehat{Z}_+ \cup \widehat{Z}_- \to \widehat{Y}. \tag{17.7}$$

On $\widehat{Z}_\pm$ we define

$$\widehat{\Omega}(x, y, z, P) = \frac{1}{2 - P}\left((x - y, -y, z, P) \pm (-2 + P, 1, 1 - P, 0)\right). \tag{17.8}$$

Lemma 17.2 $\widehat{\Omega}$ *induces a (piecewise continuous) map* $\Omega : X \to Y$ *which agrees with a projective transformation on the interior of each prism quotient.*

Proof: We have covering maps $\pi_X : \widehat{X} \to X$ and $\pi_Y : \widehat{Y} \to Y$. We define

$$\Omega = \pi_Y \circ \widehat{\Omega} \circ \pi_X^{-1}. \tag{17.9}$$

Since π_X^{-1} is multi-valued, we will check that Ω is well defined.

Let Λ_Y denote the covering group associated to $\widehat{Y}$ and Y. Let $\widehat{Z}$ stand for any of our prisms. Two inverse images of π_X in $\widehat{Z}$ differ by the vector $(0, 0, 2m, 0)$ for some integer m. We compute

$$\widehat{\Omega}(x, y, z + 2m, P) - \widehat{\Omega}(x, y, z, P) = m(0, 0, A + 1, 0).$$

Since the map $(x, y, z, A) \to (x, y, z + A, A)$ belongs to the group Λ_Y, we see that $\widehat{\Omega}$ agrees on these two points mod Λ_Y. Hence Ω is well defined. By construction the restriction of Ω to each of Z_+ and Z_- is a projective transformation which maps the slice X_P into the slice Y_A. $\square$

17.4 CHARACTERIZING THE IMAGE

In this section we characterize the image of Y.

Lemma 17.3 *The following is true.*

1. *$\Omega(Z)$ contains an open dense subset of Y.*

2. *$\Omega : Z \to Y$ is at most 2-to-1.*

3. *$\Omega(Z)$ contains every well-defined F_Y-orbit in Y.*

Proof: We will describe the results of a straightforward calculation whose details we omit. The image

$$\widehat{W}_\pm = \widehat{\Omega}(\widehat{Z}_\pm)$$

is another prism. Slicing W_+ at any Z-value and at $A = 0$ we get the square $[-1, 0] \times [0, 1]$. Slicing $\widehat{W}_+$ at any Z-value and at $A = 1$ we get the rectangle $[-1, 0] \times [-1, 2]$. Hence, the A slice of $\widehat{W}_+$ contains $[-1, 0] \times [0, 1 + A]^2$. Similarly, the A slice of W_- contains $[0, 1] \times [1 + A]^2$. Hence, the union $\widehat{W}_+ \cup \widehat{W}_-$ contains the fundamental domain

$$\bigcup_{A \in [0,1]} [-1, 1] \times [1 + A]^2$$

for the action of the group Λ_Y on the universal cover $\widehat{Y}$ of Y. This proves Statement 1.

At the same time, $\Omega(Z)$ is contained in a union of 2 fundamental domains for the action of Λ_Y. Hence $\Omega : Z \to Y$ is at most 2-to-1. This proves Statement 2.

The only points in Y that do not lie in $\Omega(X)$ have lifts to $\widehat{Y}$ with integer first coordinate. Inspecting our partition for Y, we see that these points do not have well-defined orbits. This proves Statement 3. □

Remark: In view of the structure above, we see that when $\zeta, \zeta' \in Z$ and $\Omega(\zeta) = \omega(\zeta')$, we have ζ, ζ' in the same component of Z and furthermore $\zeta - \zeta' = \pm(2, 2, 2P + 2k, 0)$ for some $k \in \mathbf{Z}$.

Computer Tie-In: On the main program, open the *plaid PET* window and the *graph PET* window at the same time. Whenever you click/drag the middle mouse button on the *plaid PET* window, you will see the corresponding Ω-image in the *graph PET* window. If you set the *slice* to Z on the *graph PET slicer* you can see that the points with integer first coordinate lie in the boundaries of the polytopes in the partition.

17.5 THE CLEAN PARTITION

Suppose that $\mathcal{A} = \{A_i\}$ and $\mathcal{B} = \{B_j\}$ are two partitions of X into polytopes. We define the *common refinement* to be

$$\mathcal{A}\#\mathcal{B} = \bigcup_{i,j} A_i \cap B_j. \tag{17.10}$$

This is the usual definition. In case the polytopes in the two partitions have rational vertices, the polytopes in the common refinement will also have rational vertices. This construction may be iterated, so that we can take the n-fold common refinement of n partitions.

The set $\widehat{Z}_\pm$ is an infinite prism. The polytope

$$Z_\pm^\bullet = \widehat{Z}_\pm \cap \left(\mathbf{R}^2 \times [-1, 1] \times [0, 1]\right) \tag{17.11}$$

is a fundamental domain for the action of $T_3 \in \Lambda_X$ on $\widehat{Z}_\pm$. The compact polytopes $Z_\pm^\bullet$ are useful for our purposes. Let $Z^\bullet = Z_+^\bullet \cup Z_-^\bullet$.

In our definitions, we use the notation $\mathcal{X}$ to denote both the checkerboard partition of X and its lift to $\widehat{X}$. We define

$$\mathcal{X}^\bullet = \mathcal{X}\#Z^\bullet. \tag{17.12}$$

Here $\mathcal{X}^\bullet$ is a partition of $Z^\bullet$ into polytopes. In this case, we are lifting to $\widehat{X}$ and then chopping off everything outside $Z^\bullet$. Only a few polytopes actually cross the boundary of $Z^\bullet$.

Next, let

$$\mathcal{Z} = \Lambda_X(\widehat{Z}_+ \cup \widehat{Z}_- \cup \widehat{Z}'_+ \cup \widehat{Z}'_-). \tag{17.13}$$

This is our tiling by prisms.

Finally, define

$$\mathcal{C} = \mathcal{X}^\bullet \,\#\, F_X^{-1}(\mathcal{X}) \,\#\, F_X^{-2}(\mathcal{Z}). \tag{17.14}$$

We call $\mathcal{C}$ the *clean partition* for reasons we will discuss momentarily. $\mathcal{C}$ is a partition of $Z^\bullet$ into 190 polytopes. It turns out that if τ is a polytope in $\mathcal{C}$

then 60τ has integer vertices.

Remark: We compute the partition $\mathcal{C}$ numerically, using non-rigorous computer programs. For each simplex τ in the computed partition, we notice that the vertices of 60τ are extremely close to integers. We then replace 60τ by the corresponding integer polytope $60\tau'$ and then set $\tau = \tau'$. The probability that $\mathcal{C}$ does not satisfy Equation 17.14 is about the same as that of an orange cow wandering into the room as I write this, but still I did not formally check that $\mathcal{C}$ satisfies Equation 17.14. Rather than perform this painful check, we simply verify directly that $\mathcal{C}$ has enough good properties to prove the Orbit Equivalence Theorem. See Lemma 17.5 below and also Step 2 in §17.8. If we knew formally that $\mathcal{C}$ really did satisfy Equation 17.14 then some of our computational checks would be redundant.

Computer Tie-In: On the main program, open up the *plaid PET* window and the *plaid PET* control panel. Turn off all the displays except the *clean partition* and you can see $\mathcal{C}$. You can also turn on the displays for $\mathcal{Z}_+$, etc. to see how $\mathcal{C}$ wits with respect to $\mathcal{Z}$. Likewise, you can see how $\mathcal{C}$ sits with respect to $\mathcal{X}$ and to $F_X^{\pm 1}(\mathcal{X})$.

17.6 THE MAIN PROOF

Recall that the map F_X on X is defined in terms of the partition $\mathcal{X}$ discussed in the previous section. Likewise, the map F_Y on Y is defined in terms of the partition defined in §13.5. We now call this partition $\mathcal{Y}$. We say that an open polytope τ is $\mathcal{X}$-*clean* if it is contained in a single polytope of $\mathcal{X}$. We make similar definitions for $\mathcal{Y}$ and $\mathcal{Z}$. When τ is clean with respect to one of the partitions, the associated map is entirely defined on τ and also the restriction of a projective transformation. (Affine maps count as projective transformations.)

Lemma 17.4 (Computer Assist 1) *The set Z is dynamically large.*

Proof: We check that every polytope of $\mathcal{X}$ is $\mathcal{Z}$-clean. Hence, the interior of each polytope of $\mathcal{X}$ either lies in the orbit $\Lambda_X(\widehat{Z})$ or in the orbit $\Lambda_X(\widehat{Z}')$. We call the polytopes of $\mathcal{X}$ *good* and *bad* accordingly. It turns out that there are 42 good polytopes and 10 bad polytopes modulo the action of Λ_X. We choose representatives $\tau_1, ..., \tau_{10}$ of the bad polytopes. We check that $F_X(\tau_i) \in Z$ for $i = 1, ..., 10$. $\square$

Now we turn to the task of proving Equation 17.1. We establish the following result computationally in the sections following this one.

Lemma 17.5 (Computer Assist 2) $\mathcal{C}$ *is a partition of $Z_+^{\bullet}$ and for any polytope τ of $\mathcal{C}$ the following is true.*

1. $F_X^k(\tau)$ is $\mathcal{X}$-clean for $k = 0, 1$.

2. $\Omega(\tau)$ is $\mathcal{Y}$-clean.

3. $F_X^k(\tau)$ is $\mathcal{Z}$-clean for $k = 0, 1, 2$.

For any polytope $\tau \in \mathcal{C}$ and either $k \in \{1, 2\}$ both maps $\Omega \circ F_X^k$ and $F_Y \circ \Omega$ are entirely defined on the interior of τ. Moreover, these maps are projective transformations on the interior of τ whose restriction to each slice

$$\tau_P = \tau \cap (\mathbf{R}^3 \times \{P\})$$

is affine. We extend these maps to the closure of τ in the unique way that keeps them continuous.

Lemma 17.6 (Computer Assist 3) *For each $\tau \in \mathcal{C}$ there is some $k \in \{1, 2\}$ such that $f_1 = \Omega \circ F_X^k$ and $f_2 = F_Y \circ \Omega$ agree on the interior of τ.*

Proof: To highlight the logic of the proof, we let $\overline{f}_j$ denote the continuous extension of f_j to the closure $\overline{\tau}$ of τ. We check by direct calculation that $\overline{f}_1$ and $\overline{f}_2$ agree on the vertices of $\overline{\tau}$. Let e be an edge of $\overline{\tau}$ that connects two vertices of $\overline{\tau}$ having different P-coordinate. We know that $\overline{f}_1$ and $\overline{f}_2$ both have the same action on the endpoints of e. Moreover, both maps carry the point e_P into $\widehat{Y}_A$. Hence $\overline{f}_1 = \overline{f}_2$ on e. But then $\overline{f}_1 = \overline{f}_2$ on the vertices of $\overline{\tau}_P$. An affine transformation is uniquely determined by where it sends the vertices of a polygon. Hence $\overline{f}_1 = \overline{f}_2$ on $\overline{\tau}_P$. Since this is true for each P, we see that $\overline{f}_1 = \overline{f}_2$ on $\overline{\tau}$. Hence $f_1 = f_2$ on the interior of τ. $\square$

Lemma 17.7 (Computer Assist 4) *For each $\tau \in \mathcal{C}$ such that $F_X(\tau) \subset Z$, the maps Ω and $\Omega \circ F_X$ agree on the interior of τ. Furthermore, F_X is the identity on τ if and only if F_Y is the identity on $\Omega(\tau)$.*

Proof: We prove this in the same way as Lemma 17.6. $\square$

There is an open dense subset of points $\zeta \in Z$ such that ζ lies in the interior of a polytope of $\mathcal{C}$. Lemmas 17.6 and 17.7 immediately show that Statements 1 and 2 of the Orbit Equivalence Theorem hold for such ζ. If these statements ever fail, then they fail on an open set, and this contradicts what we have already proven. Finally, we observe, from the formulas for F_X given in §8.5, that if F_X is entirely defined on a polytope τ, then F_X either fixes every point of τ or nontrivially moves every point of τ. The same goes for F_Y on $\Omega(\tau)$. So, by Lemma 17.6, a point $\zeta \in X$ is fixed by F_X if and only if $\Omega(\zeta)$ is fixed by F_Y. This establishes Statement 3 of the Orbit Equivalence Theorem.

This proves the Orbit Equivalence Theorem modulo the 4 computer-assisted lemmas.

17.7 COMPUTATIONAL TECHNIQUES

Now we build the computational framework for the computer-assisted lemmas in the previous section.

Tight Polytopes: Say that a *tight polytope* is a convex polytope in $\mathbf{R}^4$ with integer vertices, such that each vertex is the unique extreme point of some linear functional. In ther words, a tight polytope is the convex hull of its (integer) vertices, and the convex hull of any proper subset of vertices is a proper subset. We always deal with tight polytopes. The polytopes in our partitions have all the properties mentioned above, except that their vertices are rational rather than integral. We fix this problem by scaling all polytopes in all partitions by a factor of 60. This makes them all convex integral tight polytopes.

Tight Polytope Test: Suppose we are given a finite number of integer points in $\mathbf{R}^4$. Here is how we test that they are the vertices of a tight polytope. We consider all linear functionals of the form

$$L(x, y, z, A) = c_1 x + c_2 y + c_3 z + c_4 A, \qquad |c_i| \leq N \qquad (17.15)$$

and we wait until we have shown that each vertex is the unique maximum for one of the functionals. For the polytopes of interest to us, it suffices to take $N = 5$. In general, our test halts with success for some N if and only if the polytope is tight.

Disjointness Test: Here is how we verify that two tight polytopes P_1 and P_2 have disjoint interiors. We consider the same linear functionals as listed in Equation 17.15 and we try to find some such L with the property that

$$\max_{v \in V(P_1)} L(v) \leq \min_{v \in V(P_2)} L(v). \qquad (17.16)$$

Here $V(P_k)$ denotes the vertex set of P_k. If this happens, then we have found a hyperplane which separates the one polytope from the other. This time we take $N = 5$.

Containment Test: Given tight polytopes P_1 and P_2, here is how we verify that $P_1 \subset P_2$. By convexity, it suffices to prove that $v \in P_2$ for each vertex of P_1. So, we explain how we verify that $P = P_2$ contains an integer point v. We do not have the explicit facet structure of P, though for another purpose (computing volumes) we do find it.

Let $\{L_k\}$ denote the set of all linear functionals determined by 4-tuples of vertices of P. Precisely, Given 4 vertices of P, say w_0, w_1, w_2, w_3, and some integer point v, we take the 4×4 matrix whose first three rows are $w_i - w_0$ for $i = 1, 2, 3$ and whose last row is v. Then

$$L_{w_0, w_1, w_2, w_3}(v) = \det(M) \qquad (17.17)$$

is the linear functional we have in mind.

If the vertices do not span a 3-dimensional space, then L will be trivial. This does not bother us. Also, some choices of vertices will not lead to linear functions which define a face of P. This does not bother us either. The point is that our list of linear functionals contains all the ones which do in fact define faces of P. We take our vertices in all orders, to make sure that we pick up every possible relevant linear functional. The computer does not mind this redundancy.

It is an elementary exercise to show that $v \notin P$ if and only if v is a unique extreme point amongst the set $\{v\} \cup V(P)$ for one of our linear functionals. Here $V(P)$ denotes the vertex set of P, as above. So, $v \in P$ if and only if v is never a unique extreme point for any of the linear functionals on our list.

Volume: First we explain how we find the codimension 1 faces of P. We search for k-tuples of vertices which are simultaneously in general position and the common extreme points for one of the linear functionals on our list. As long as $k \geq 4$, the list we find will be the vertices of one of the faces of P.

Now we explain how we compute the volume of P. This is a recursive problem. Let v_0 be the first vertex of P. Let $F_1, ..., F_k$ be the codimension 1 faces of P. Let P_j be the cone of F_j to v_0. This is the same as the convex hull of $F_j \cup \{v_0\}$. Then

$$\mathrm{vol}(V) = \sum_{j=1}^{k} \mathrm{vol}(V_j). \tag{17.18}$$

If $v_0 \in V_j$ then the volume is 0. These extra trivial sums do not bother us.

To compute $\mathrm{vol}(V_j)$ we let $w_{j0}, w_{j1}, w_{j2}, w_{j3}$ be the first 4 vertices of F_j. Let L_j be the associated linear functional. Then

$$4 \times \mathrm{vol}(V_j) = L_j(v_0 - w_{j0}) \times \mathrm{vol}(F_j). \tag{17.19}$$

We compute $\mathrm{vol}(F_j)$ using the same method, one dimension down. That is, we cone all the facets of F_j to the point w_{j0}. It turns out that the polyhedra $\{F_{ij}\}$ in the subdivision of F are either tetrahedra or pyramids with quadrilateral base. In case F_{ij} is a tetrahedron we compute $12\mathrm{vol}(F_{ij})$ by taking the appropriate determinant and doubling the answer. In the other case, we compute 6 times the volume of each of the 4 sub-tetrahedron of F_{ij} obtained by omitting a vertex other than w_{0j} and then we add up these volumes. This computes $12\mathrm{vol}(F_{ij})$ regardless of the cyclic ordering of the vertices around the base of F_{ij}.

When we add up all these contributions, we get $12 \ \mathrm{vol}(F_j)$. So, our final answer is $48 \ \mathrm{vol}(V)$. The reason we scale things up is that we want to have entirely integer quantities.

Controlled Division: Since we only deal with integers, we have to take special care when computing a projective transformation. Every time we perform the operation a/b we first check that $a \equiv 0 \mod b$. When this happens, we call the division operation *legal*. Otherwse, we call it *illegal*. We get lucky, and every division we need to make is legal. The canonical nature of what we are doing underlies this luck, but I don't have a complete explanation. Here is a partial explanation: The map Ω carries the P slices to the A slices. When we scale up by 60, the last coordinate of the points in the domains for Ω is $60P$ and the last coordinate in the range is $60A$. When $A = p/q$ we have $p = 2p/(p+q)$. Hence, the denominator of A is less than the denominator of P. We only compute Ω at points where the denominator of P is in the set $\{1, 2, 3, 5\}$. This forces the denominator for A to lie in the set $\{1, 2, 3, 4\}$. But then $60A$ is an integer.

Here is an example of our luck in action. One of the polytopes τ in the clean partition, when scaled by 60 has vertices which are the rows of the following matrix:

$$\begin{bmatrix} 180 & 60 & -60 & 0 \\ 132 & 60 & 12 & 24 \\ 60 & 60 & -60 & 0 \\ 60 & -60 & -60 & 0 \\ 120 & 60 & 0 & 30 \end{bmatrix}.$$

The 60-scaled version of $\Omega(\tau)$ has vertices which are the rows of the following matrix:

$$\begin{bmatrix} 0 & 0 & -60 & 0 \\ -15 & 0 & -45 & 15 \\ -60 & 0 & -60 & 0 \\ 0 & 60 & -60 & 0 \\ -20 & 0 & -60 & 20 \end{bmatrix}.$$

Overflow Error: The only danger in performing integer calculations on the computer is overflow error. The integers we use might be too large. We represent integers by *longs* in java and this gives us about 16 digits. Our calculations come nowhere near these limits. The largest calculation, by far, computes a volume of 44561920000.

17.8 THE CALCULATIONS

We have put all the routines needed for the proofs in a separate directory, **/Proofs:**. The needed routines are contained in a small subset of the files, which the reader can probably inspect. The files beginning with **Data** store the integer coordinates and the dynamical information for each of the polytopes of interest to us. The vertices for $\mathcal{X}$ are listed in §8.3. The vertices for $\mathcal{Y}$ are listed in §13.6. The reader will need to inspect the files to see the vertices for the polytopes in $\mathcal{C}$ and $\mathcal{Z}$. By *dynamical indication* we mean a short combinatorial code which explains which branch of the relevant PET acts on the polytope.

We prove the computer-assisted lemmas in 6 steps.

Step 1: We check that every polytope of each of the 4 partitions

$$\{\mathcal{X}, \mathcal{Y}, \mathcal{Z}, \mathcal{C}\}$$

is tight. This allows us to use all our methods on every polytope in sight.

Step 2: We check that $\mathcal{C}$ really is a partition of the polytope $Z^\bullet$. By symmetry, it suffices to check this for the partition $\mathcal{C}_+$ of $Z^\bullet_+$. The other half of the partition is obtained by applying the map I. We accomplish our goal with three calculations, as follows:

- We check that the sum of the volumes of the polytopes in $\mathcal{C}_+$ equals the volume of $Z^\bullet_+$. The common volume is $456192000 = 60^4 \times (176/5)$.

- We check that the polytopes in $\mathcal{C}_+$ have pairwise disjoint interiors.

- We check that every polytope in $\mathcal{C}_+$ is contained in $Z^\bullet_+$.

These three facts together imply that $\mathcal{C}_+$ is indeed a partition of $Z^\bullet_+$.

Step 3: We perform Computer Assist 2 mentioned above. For each polytope τ of interest to us, we verify an equation of the form $f(\tau) \subset \lambda(\sigma)$. Here

- f is the map of interest to us. Here $f \in \{\text{Identity}, F_X, F_X^2, \Omega\}$ depending on the case.

- σ is one of the stored polytopes in the relevant partition. The partition is one of $\mathcal{X}$, $\mathcal{Y}$, or $\mathcal{Z}$, depending on the case.

- λ is a member of the group of interest to us, either Λ'_X or Λ_Y, depending on the case. The element λ always has the form $R^r S^s T^t$ where R, S, T are the generators and $\max(|r|, |s|, |t|) = 3$.

By Λ'_X we mean the group from §8.3 that contains Λ_X with index 2. We use Λ'_X in place of Λ_X because we only store coordinates for half the polytopes mod Λ_X. The other half are obtained from these by applying the map $T_1(x, y, z, P) = (x + 2, y + P, y + P, P)$.

Step 4: Step 3 contains the calculation that every $\mathcal{X}$ polytope is $\mathcal{Z}$ clean. It turns out that 5 out of the 26 stored $\mathcal{X}$ polytopes, namely those indexed $9, 10, 11, 12, 13$, lie in Z' rather than in Z. These represent half the 10 bad polytopes mentioned in Computer Assist 1 above. The other half are obtained from these 5 by applying the generator T_1 of Λ'_X.

For τ each of the 5 polytopes, we check that $F_X^{\pm 1}(\tau) \subset Z$. By symmetry, this implies that F_X maps each of the 10 bad polytopes from Computer Assist 1 into Z, as desired.

Step 5: We perform Computer Assist 3 from the previous section. That is, we check Equation 17.1 on each polytope τ of $\mathcal{C}$. We store in advance the number $k = k(\tau) \in \{1, 2\}$ and then check that the two polytopes $\widehat{\Omega} \circ F_X^k(\tau)$ and $F_Y \circ \widehat{\Omega}(\tau)$ are the same labeled polytope modulo the action of Λ_Y. In practice, we do what we did in Step 3. This check implies that

$$\Omega \circ F_X^k(\tau) = F_Y \circ \Omega(\tau)$$

in the strong sense that the ith vertex of one matches the ith vertex of the other for all i. We include in our verification the check that $F_X^k(\tau) \subset Z$.

Step 6: We perform Computer Assist 4 from the previous section. That is, we check Equation 1 on each polytope τ of $\mathcal{C}$ such that $F_X(\tau) \subset Z$. We also check that F_X fixes the vertices of τ if and only if F_Y fixes the vertices of $\Omega(\tau)$. The method of checking is the same as in Step 5.

The calculations ran to completion in about 146 seconds when I ran them on my MacBook Pro on November 10, 2017. This completes the proof of the Orbit Equivalence Theorem.

Chapter Eighteen

The Quasi-Isomorphism Theorem

18.1 CHAPTER OVERVIEW

In this chapter we prove the Quasi-Isomorphism Theorem modulo two technical lemmas which we deal with in the next two chapters. For convenience, we restate the result. Recall that Γ_A is the arithmetic graph at the parameter A, the union of the graph polygons, and PL_A is the plaid model at parameter A, the union of the plaid polygons.

Theorem 18.1 (Quasi-Isomorphism) *For each even rational parameter A, there exists an affine transformation $T_A : \mathbf{R}^2 \to \mathbf{R}^2$, and a bijection between the components of $T_A(\Gamma_A)$ and the components of PL_A with the following property. If $\gamma \in T_A(\Gamma_A)$ and $\pi \in PL_A$ are corresponding components then there is a homeomorphism $h : \gamma \to \pi$ which moves points by no more than 2 units.*

In §18.2 we will introduce the affine transformation T_A from the Quasi-Isomorphism Theorem.

In §18.3 we will define the *graph grid* $G_A = T_A(\mathbf{Z}^2)$ and state the Grid Geometry Lemma, a result about the basic geometric properties of G_A. We prove the Grid Geometry Lemma in §19.

In §18.4 we introduce the set Z^* that appears in the Renormalization Theorem and state the main result about it, the Intertwining Lemma. We will prove the Intertwining Lemma in §20.

In §18.5 we will explain how the Orbit Equivalence Theorem sets up a canonical bijection between the nontrivial orbits of the plaid PET and the orbits of the graph PET.

In §18.6 we will reinterpret the orbit correspondence in terms of the plaid polygons and the arithmetic graph polygons. We then put everything together and complete the proof of the Quasi-Isomorphism Theorem.

In §18.7 we deduce the Projection Theorem (Theorem 0.2) from the Quasi-Isomorphism Theorem.

Finally, in §18.8 we will give a sketch of the connection between the Quasi-Isomorphism Theorem and renormalization.

18.2 THE CANONICAL AFFINE TRANSFORMATION

Let Λ_Y be the graph PET group. Let $A = p/q$ be an even rational parameter. We use the notation from §1.2. Set $\theta = 1$ when $\widehat{\tau}$ is even and $\theta = 0$ when $\widehat{\tau}$ is odd. We define

$$T_A \begin{pmatrix} m \\ n \end{pmatrix} = \frac{1}{A+1} \begin{pmatrix} A^2 + A & A+1 \\ -A^2 + 2A + 1 & -2A \end{pmatrix} \begin{pmatrix} m \\ n \end{pmatrix} +$$

$$\begin{pmatrix} \frac{1}{2q} \\ \frac{-1}{2q} + \frac{1}{\omega} + \widehat{\tau} + \omega\theta \end{pmatrix}. \tag{18.1}$$

Remark: I found the linear part of this map by trying to simplify the slopes of the grid lines from the Hexagrid Theorem in [**S1**]. Then I fiddled around with the translation part experimentally until it came out exactly right.

The linear part dT_A of T_A is defined for irrational parameters as well as rational parameters, but the map itself is only defined when $A = p/q$ is an even rational. Let Ψ'_A be the map from §13.5, namely

$$\Psi'_A(m,n) = (t+t', t-t', t-t', A) \bmod \Lambda_Y, \quad t = Am+n+\frac{1}{2q}, \quad t' = m-n. \tag{18.2}$$

Define

$$\Psi_A = \Psi'_A \circ T_A^{-1}. \tag{18.3}$$

The Graph Master Picture Theorem remains true with $T_A(\Gamma_A)$ replacing Γ and Ψ_A replacing Ψ'_A and $G_A = T_A(\mathbf{Z}^2)$ replacing $\mathbf{Z}^2$. Henceforth, this is the form of the Graph Master Picture Theorem we use.

A calculation, which we omit, shows that

$$\Psi_A(x,y) = (x+\mu, x-\mu, x-\mu, A) \bmod \Lambda_Y,$$

$$\mu(x,y) = 1 + \frac{x(A-1) + y(A+1)}{A+1}. \tag{18.4}$$

The linear part of this equation is easy to see, and the translation part is somewhat more tedious to derive. We emphasize that this equation is only meant to work on G_A. This nice equation partially explains the canonical nature of the Canonical Affine Transformation.

In [**S1**, §12] we work out the translation symmetries of the arithmetic graph. In the oriented case, these symmetries form a lattice generated by the vectors $(2q, -2p)$ and $(0, \omega^2)$. The map T_A maps these vectors to $(0, 2\omega)$ and $(\omega^2, -2p\omega)$. These latter two vectors generate the symmetry lattice of the oriented plaid model. In short, T_A maps the translation symmetry lattice of the oriented arithmetic graph Γ_A to the translation symmetry lattice of the oriented plaid model PL_A.

Computer Tie-In: On the main program open the *debug* control panel and select the *AG transforms* option. Press *go* and the program will check numerically on random inputs that Equation 18.4 is correct. This isn't really a visual display of anything. We mention it just to let the reader know that we did check this equation numerically.

18.3 THE GRAPH GRID

We have $G_A = T_A(\mathbf{Z}^2)$. We call G_A the *graph grid*. We emphasize that G_A is not a lattice in $\mathbf{R}^2$ because $(0,0) \notin G_A$. However, it turns out that G_A always has rotational symmetry about $(0,0)$. This fact allows us to give a more robust definition. We redefine G_A to be the translate of the lattice $dT_A(\mathbf{Z}^2)$ which does not contain the origin but which has the origin as a point of rotational symmetry. This defines G_A for all $A \in [0,1]$ and agrees with the initial definition. As A varies from 0 to 1, the grid G_A interpolates between the grid of half-integers and the grid of integers whose coordinates have odd sum, and each individual point travels along a hyperbola or straight line.

Computer Tie-In: On the main program, open up the *planar window* and the *arithmetic graph* control panel. Turn on the *polygons* option on the *arithmetic graph* control panel. Assuming that you have just opened the program, the planer window will show the plaid polygons and the normalized graph superimposed over each other. You can see pictures like Figure 0.5 for any (smallish) parameter you like. To rapidly sample different parameters, click the *random* key repeatedly on the *parameter entry window* at top right.

Next, open up the *grid analyzer window*. Two windows will pop up. One of the windows plots the graph grid and the square grid superimposed on top of each other. There is a slider at the top of this window that lets you control the parameter A. If you click the middle mouse button (or use key-x) on a square in this window, you will see a yellow arc that indicates the motion of the nearest grid point as a function of the parameter. This yellow curve is the line/hyperbola just mentioned.

We fix some even rational parameter $A = p/q$ throughout this section. For notational convenience we sometimes suppress A from our notation. A *distinguished edge* in the graph grid is one connecting distinct points of the form $T(\zeta)$ and $T(\zeta + (i,j))$ for $i,j \in \{-1,0,1\}$.

Lemma 18.2 (Grid Geometry) *The following is true at each even rational parameter.*

1. *No point of the graph grid lies on the boundary of a square.*

2. *Two points of the graph grid cannot lie in the same square.*

3. *A distinguished edge has length less than* 2.

4. *A union of* 2 *vertically consecutive squares intersects the graph grid.*

5. *A union of* 3 *horizontally consecutive squares intersects the graph grid.*

Proof: See §19. □

188 CHAPTER 18

Let Π denote the plaid grid. This is the set of points in $\boldsymbol{R}^2$ having half-integer coordinates. We say that $c \in \Pi$ is *grid full* if there is a point $g \in G_A$ which lies in the unit integer square centered at c. Thanks to the Grid Geometry Lemma, g is unique provided that g exists. We write $c \to_A g$. We let Π_A denote the set of grid full points in Π.

For the purposes of working with plaid polygons, we re-draw them so that the line segment within each tile is replaced by a possibly broken line segment which contains the center of the tile. Figure 18.1 shows what we mean. By the *vertices* of a plaid polygon we mean the points of the plaid grid, as highlighted on the right-hand side of Figure 18.1.

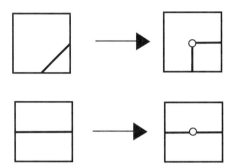

Figure 18.1: New style of drawing the plaid polygons.

Lemma 18.3 *Let A be any even rational parameter and let π be some plaid polygon. Any set of 3 consecutive vertices of π contains a point of Π_A. In particular, some vertex of π lies in Π_A.*

Proof: Any arc of π connecting 3 consecutive vertices either contains a vertical segment or two consecutive horizontal segments. Either way, the arc intersects Π_A, by the Grid Geometry Lemma. $\square$

18.4 THE INTERTWINING LEMMA

Let Z be the set from the Orbit Equivalence Theorem. Here we define a smaller set $Z^* \subset Z$ which is still dynamically large.

Just as $Z = Z_+ \cup Z_-$, with $Z_- = I(Z_+)$ and

$$I(x, y, z, P) = (-x, -y, -z, P),$$

we have $Z^* = Z_+^* \cup Z_-^*$ with $Z_-^* = I(Z_+^*)$. We define

$$Z_+^* = \bigcup_{P \in [0,1]} Z_{P,+}^*, \qquad Z_P^* = H_{P,+}^* \times \boldsymbol{R}. \tag{18.5}$$

Here $H_{P,+}^*$ is the parallelogram with vertices

$$(P-1, P-1), \ (1-P, -1), \ (3-P, 1), \ (P+1, P+1).$$

The darkly shaded parallelogram in Figure 18.2 is $H^*_{P,+}$. This is a (Z, P) slice of Z^*_+. The slice is a parallelogram, but the slope of the top and bottom sides, namely $P/(2P - 2)$, varies with P. This makes it impossible for Z^*_+ to be a convex polytope. Figure 18.2 also shows the larger set Z_+ for the sake of comparison.

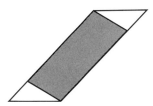

Figure 18.2: (Z, P) slices of Z^*_+ and Z_+.

Let Ω be the map from the Orbit Equivalence Theorem. Recall that $\Phi_A : \Pi \to X_P$ and $\Psi_A : G_A \to Y_A$ are the maps which appear respectively in the Plaid Master Picture Theorem and the Graph Master Picture Theorem. Here is our result which links all these maps together.

Lemma 18.4 (Intertwining) *Let A be an even rational parameter and let $P = 2A/(1 + A)$. Then*

1. $\Phi_A(\Pi_A) \subset Z^*_P$.

2. $\Omega(\Phi_A(c)) = \Psi_A(g)$ *when* $c \in \Pi_A$ *and* $g \in G_A$ *and* $c \to_A g$.

3. Ω *is an injective map from (the interior of) Z^* into Y.*

Proof: See §20. $\square$

18.5 THE CORRESPONDENCE OF ORBITS

We fix some even rational parameter A and let $P = 2A/(1 + A)$ as usual. We use the notation from the Orbit Equivalence Theorem. We have the plaid PET $F_X : X \to X$ and the graph PET $F_Y : Y \to Y$, and our map $\Omega : Z \to Y$.

Let $\{x_i\}$ be a nontrivial F_X-orbit with respect to the parameter P. Since P is rational, this orbit is necessarily finite. Since Z is dynamically large, there is some index i such that $x_i \in Z$. We associate to $\{x_i\}$ the F_Y-orbit of $\Omega(x_i)$.

Lemma 18.5 *The orbit correspondence is well defined.*

Proof: Suppose that i and j are two indices such that $x_i, x_j \in Z$. We want to show that the F_Y-orbit of $\Omega(x_i)$ coincides with the F_Y-orbit of $\Omega(x_j)$. Call

i and j *neighbors* if, in the cyclic ordering of the indices, we have $i < j \le i+2$ and there is no index $k \in (i, j)$ with $x_k \in Z$.

The Orbit Equivalence Theorem implies that any two indices i, j can be joined in a chain $i = i_0, ..., i_n = j$ where each consecutive pair consists of neighbors. So, it suffices to prove our result when i and j are neighbors. Let $y_j = \Omega(x_j)$. There are two cases. If $j = i + 1$ then $x_j = F_X(x_i)$. In this case, either $y_j = y_i$ or else $y_j = F_Y(y_j)$ according to whether $k(x_i) = 1$ or $k(x_i) = 2$ in the Orbit Equivalence Theorem. If $j = i + 2$ then $k(x_i) = 2$ and $y_j = F_Y^2(y_i)$. This completes the proof. □

Here is a different way to summarize the analysis in the lemma above. If $\{y_i\}$ is the F_Y-orbit corresponding to the F_X-orbit $\{x_i\}$ then

$$\{y_j\} = \Omega(\{x_i\} \cap Z). \tag{18.6}$$

Let O_X denote the set of nontrivial F_X-orbits and let O_Y denote the set of nontrivial F_Y orbits. Our correspondence gives a well-defined map $O_X \to O_Y$. Since $\Omega(Z)$ contains all the well-defined F_Y-orbits, our correspondence is surjective. Since $\Omega : Z \to Y$ is at most 2-to-1, Equation 18.6 tells us that our correspondence is at most 2-to-1.

We say that a *special point* in the plaid PET $F_X : X \to X$ is a point of the form $\Psi_A(\zeta)$ for an even rational parameter A and some point $\zeta \in \Pi$ in the plaid grid. By the Plaid Master Picture Theorem, the set of special points is invariant under the action of F_X. A *special orbit* in X is the orbit of a special point. Let O_X^* denote the set of special orbits in X.

We make the same definition of special points and orbits in $F_Y : Y \to Y$ relative to the graph grid G_A and the map Ψ_A. Let O_Y^* denote the special orbits in Y.

Lemma 18.6 *The orbit correspondence $O_X \to O_Y$ induces a surjective correspondence $O_X^* \to O_Y^*$.*

Proof: Let $\{x_i\} \subset X$ be a special orbit. By Lemma 18.3 we know that some $x = x_i$ has the form $x = \Phi_A(c)$ where $x \in \Pi_A$. Let $g \in G_A$ be such that $c \to_A g$ and let $y = \Psi_A(c)$. By the Intertwining Lemma, $\Omega(x) = y$. But then the (special) orbit $\{y_i\}$ of $y \in Y$ corresponds to $\{x_i\}$. This shows that our correspondence $O_X \to O_Y$ maps O_X^* into O_Y^*.

The fact that the correspondence $O_X \to O_Y$ is surjective does not quite prove that the correspondence $O_X^* \to O_Y^*$ is surjective, so we give a proof. Given any special F_Y-orbit $\{y_i\}$ let $y = y_1$. Let $g \in G_A$ be such that $\Psi_A(g) = y_1$. Let $c \in \Pi_A$ be such that $c \to_A g$. Let $x = \Phi_A(c)$. By the Intertwining Lemma, $\Omega(x) = y$. Hence $\{y_i\}$ corresponds to the orbit of x. This shows that the correspondence $O_X^* \to O_Y^*$ is surjective. □

Our next goal is to show that our correspondence $O_X^* \to O_Y^*$ is a bijection. We first prove a helpful technical lemma.

Lemma 18.7 *Suppose that $c, c' \in \Pi$ are points in the same plaid block such that $\Phi_A(c) \neq \Phi_A(c')$ and $\Omega \circ \Phi_A(c) = \Omega \circ \Phi_A(c')$. Then $c - c' = (0, \pm 1)$.*

Proof: We set $(m, n) = c - c'$. Let $\zeta = \Phi_A(c)$ and $\zeta' = \Phi_A(c')$. We have

$$\zeta - \zeta' = m(2P, 2P, 2P, 0) + n(2, 0, 2P, 0) \bmod \Lambda_X. \qquad (18.7)$$

At the same time, when we normalize both these images to lie in Z^*, the remark at the end of §17.4 tells us that

$$\zeta - \zeta' = \pm(2, 2, 2P + 2k, 0). \qquad (18.8)$$

Let $\omega = p + q$, where $p/q = A$ is the parameter. Looking at the first coordinate of Equation 18.7 and noting the vectors generating Λ_X have coordinates in $4\mathbf{Z}$, we have

$$2mP - x = 2mP + 2n \pm 2 = \frac{4mp}{\omega} + 2n \pm 2 \in 4\mathbf{Z}. \qquad (18.9)$$

Since p and ω are odd, this means that ω divides m. But $|m| < \omega$ because c and c' lie in the same block. Hence $m = 0$ and n is odd. The third coordinate of $\zeta - \zeta'$ is $2(n \pm 1)P + 2k$ and this must be in $2\mathbf{Z}$. The only solutions with $|n| < \omega$ are $n = \pm 1$ or $n = \pm(\omega - 1)$ and we can rule out the latter solutions because $\omega - 1$ is even. $\square$

Lemma 18.8 *The correspondence $O_X^* \to O_Y^*$ is a bijection.*

Proof: From what we know already, it suffices for us to show that the correspondence is injective. Suppose for the sake of contradiction that we have special F_X-orbits $\{x_i\}$ and $\{x_i'\}$ and a special F_Y-orbit $\{y_i\}$ such that $\{x_i\} \to \{y_i\}$ and also $\{x_i'\} \to \{y_i\}$. We have the basic equation

$$\Omega(\{x_i\} \cap Z) = \Omega(\{x_i'\} \cap Z) = \{y_i\}. \qquad (18.10)$$

Let π and π' be two plaid polygons corresponding to $\{x_i\}$ and $\{x_i'\}$. Let c_i and c_i' be the vertices of π and π' respectively corresponding to x_i and x_i'. We call c_i *friendly* if $x_i \in Z$ and otherwise *unfriendly*. We make the same definition for c_i'. We call c_i and c_j' *friends* if c_i and c_j' are friendly and if $\Omega(c_i) = \Omega(c_j')$. By equation 18.10, every friendly vertex has a friend. When c_i and c_j' are friends, Lemma 18.7 says $(c_i - c_j') = (0, \pm 1)$.

By the Orbit Equivalence Theorem, at least one out of every two consecutive vertices of π is friendly, and thus lies either directly above or directly below a vertex of π'. This would only be possible if π and π' are contained in (adjacent) horizontal lines, but that cannot happen for closed polygons. $\square$

Henceforth we consider only special orbits.

Lemma 18.9 *Suppose that $\{x_i\} \to \{y_i\}$. Then Ω induces a bijection between $\{x_i\} \cap Z^*$ and $\{y_i\}$.*

Proof: Since Ω is injective on Z^*, we see that Ω is also injective on $\{x_i\} \cap Z^*$. For each y_j there is some $g_j \in G_A$ such that $\Psi_A(g_j) = y_j$. There is some $c \in \Pi_A$ such that $c \to_A g_j$. By the Intertwining Lemma, $\Omega(x) = y_j$ where $x = \Phi_A(c)$. Hence, the map from $\{x_i\} \cap Z^*$ to $\{y_i\}$ is surjective. $\square$

18.6 THE END OF THE PROOF

Let PL_A denote the set of plaid polygons relative to the parameter A. Let Γ_A denote the set of arithmetic graph polygons relative to the parameter A. By the Plaid Master Picture Theorem, every element of PL_A corresponds to an element of O_X. For each element of O_X there are infinitely many elements of PL_A before we mod out by symmetry. Rather than mod out by symmetry, we simply call two elements of PL_A *equivalent* if they correspond to the same element of O_X. We make the same definition for Γ_A relative to O_Y. Given $\pi \in PL_A$ and $\gamma \in \Gamma_A$ we write $[\pi] \to [\gamma]$ if the special orbits $\{x_i\}$ and $\{y_j\}$ respectively corresponding to π and γ satisfy $\{x_i\} \to \{y_j\}$. Here $[\pi]$ and $[\gamma]$ are the equivalence classes of π and γ respectively.

We say that two points in $\boldsymbol{R}^2$ are *partnered* if they lie in the same unit integer square. We will also say, in these circumstances, that the one point is partnered with the other. We call a vertex of π *grid full* if it lies in Π_A.

Lemma 18.10 *Let π be a plaid polygon. Let E_π be the equivalence class of arithmetic graph polygons which correspond to π under the orbit correspondence. We can choose a unique representative $\gamma' \in E_\pi$ such that every grid full vertex of π is partnered with a vertex of γ and every vertex of γ is partnered with a grid full vertex of π. Here $\gamma = T_A(\gamma')$.*

Proof: Given any grid full vertex c of π, there is some $\gamma'_c \in E_\pi$ such that c is partnered with a vertex of γ_c. Two such polygons γ_1 and γ_2 corresponding to adjacent vertices c_1 and c_2 of π are such that γ'_1 and γ'_2 come within 3 units of each other. Since γ'_1 and γ'_2 are equivalent modulo the symmetries of the arithmetic graph, γ_1 and γ_2 are equivalent modulo the symmetries of the plaid model. This forces $\gamma_1 = \gamma_2$.

So, there is a single $\gamma' \in E_\pi$ such that every grid full vertex of π is partnered with a vertex of γ. At the same time, every vertex of γ is partnered with a vertex of some plaid polygon equivalent to π. But two such equivalent polygons cannot lie in the same block unless they coincide. Hence every vertex of γ is partnered with some grid full vertex of π. $\square$

Let π and γ' be as in the previous lemma. Let g_1 and g_2 be two consecutive vertices of γ. Let $\square_j$ be the unit integer square containing v_j. Let c_j be the center of $\square_j$. Let $\gamma[g_1, g_2]$ be the edge of γ connecting g_1 to g_2. Likewise let $\pi[c_1, c_2]$ be the arc of π connecting c_1 to c_2. Here are some metric properties of these objects.

- The distance from g_j to c_j is at most $\sqrt{2}/2$. Hence the affine isomorphism from $\gamma[g_1, g_2]$ to $\overline{c_1 c_2}$ moves points by at most $\sqrt{2}/2$.

- Considering the several possibilities given by Lemma 18.3, perpendicular retraction of the portion $\pi[c_1, c_2]$ onto the segment $\overline{c_1 c_2}$ moves points by at most $2/\sqrt{5}$. The extreme case is when $\pi[c_1, c_2]$ is a 1 by 2 path.

Composing the maps just discussed, we get a piecewise affine homeomorphism from $\gamma[g_1, g_2]$ to $\pi[c_1, c_2]$ which moves points by no more than

$$2/\sqrt{5} + \sqrt{2}/2 < 2.$$

Concatenating all these maps, we get a homeomorphism from γ to π which moves no point more than 2 units.

Remark: We have proved the Quasi-Isomorphism Theorem for the plaid polygons when they are drawn in the new style. There is a map taking the old style polygons to the new style polygons which moves points by at most $\sqrt{2}/4$. Since $2/\sqrt{5} + \sqrt{2} + \sqrt{2}/4 < 2$, our result also works for the old style.

18.7 THE PROJECTION THEOREM

In this section we use the Quasi-Isomorphism Theorem (and other facts we have collected) to deduce the Projection Theorem from the introduction.

Theorem 18.11 (Projection) *Let $A \in (0,1)$ be an even rational parameter. Modulo the vertical translations which preserve PL_A, there is a bijection between the polygons in PL_A that lie in the right half-plane and the special outer billiards orbits relative to the kite K_A. Moreover, the plaid polygon π may be (monotonically) parameterized as $\pi = \{(x_t, y_t) |\ t \in [0, N]\}$ in such a way that the point $2x_k$ lies within 3 units of the kth point of*

$$S_\pi = O_\pi \cap (\boldsymbol{R}_+ \times \{-1, 1\})$$

for all $k \in \{1, ..., N\}$. Here O_π is the special orbit associated to π and N is the number of points S_π.

Proof: Here is a summary of some facts from §13. Let Γ_A^+ denote the portion of Γ_A that lies above the line $y = -Ax$. Let $\tau(m, n) = (m, n) + 2(q, -p)$ and

$$f(m, n) = \left(2Am + 2n, (-1)^{m+n+1} \right). \tag{18.11}$$

f sets up a bijection between the polygons in Γ_A^*/τ and the special orbits. For each component of Γ_A^+ we can index the corresponding special orbit O_γ so that the kth vertex of γ corresponds to the kth point of

$$O_\gamma \cap (\boldsymbol{R}_+ \times \{-1, 1\}) \tag{18.12}$$

under the map f.

Now we drop the second coordinate. The map $(m, n) \to 2Am + 2n$ carries the kth vertex of γ to a point in $\boldsymbol{R}_+$ which is within 1 unit of the kth point in Equation 18.12. Now we change coordinates using the Canonical Affine Transformation. We have

$$T_A \circ \tau \circ T_A^{-1}(x, y) = (x, y + 2\omega), \qquad f \circ T_A^{-1}(x, y) = 2x. \qquad (18.13)$$

Call these new maps τ' and f'. The map f' sets up a bijection between components of $T_A(\Gamma_A^+)/\tau'$ and the special orbits in such a way that the kth point of $T_A(\gamma)$ lies within 1 unit of the kth point of the set in Equation 18.12.

Finally, we replace each graph polygon $\boldsymbol{T}_A(\gamma)$ with the plaid polygon π coming from the Quasi-Isomorphism Theorem. This gives us a bijection from the set of plaid polygons modulo the vertical symmetry τ' to the set of special orbits. In replacing $T_A(\gamma)$ with π, we only introduce an error of $t = 2$ units, giving a total error of at most $3 = 1 + 2$ units. $\square$

18.8 RENORMALIZATION INTERPRETATION

We proved that for all $c \in \Pi$ we have $\Phi_A(c) \in Z^*$ if $c \in \Pi_A$, the set of grid full squares. One thing we did not prove, but which is true, is the converse. Actually $\Phi_A(c) \in Z^*$ if and only if $c \in \Pi_A$. One way to prove this is to make a count and check that the number of points in $T_A(\boldsymbol{Z}^2)/L$ coincides with the number of points of $\Phi_A(\Pi) \cap Z^*$, and then to invoke the bijection result in Lemma 8.1.

Assuming that we have this additional result, we can interpret the Quasi-Isomorphism Theorem as the statement that the graph PET is the renormalization of the plaid PET with respect to the set Z^*. That is, the first return map of F_X to Z^* is conjugate, via Ω, to the map F_Y, at least when these maps are restricted to points in the image of Φ_A and Ψ_A.

This interpretation is easiest to see by considering the curve-following dynamics. Let π be a plaid polygon and let γ be the corresponding graph polygon. As we follow around π, we can simply make note of when we encounter a grid full vertex. When we move from one grid full vertex to the next, γ follows right along. Moving from grid full vertex to grid full vertex corresponds exactly to looking at the first return map of F_X to Z^*, by the Plaid Master Picture Theorem. Moving from vertex to vertex along Γ corresponds exactly to looking at the action of F_Y, by the Graph Master Picture Theorem.

The renormalization result should also be true without the restriction that we just look at the points in X and Y corresponding to the images of Φ and Ψ. The basic idea here is that every other orbit should have the same combinatorial structure as an orbit of one of these special points. We do not pursue this, however.

Chapter Nineteen

Geometry of the Graph Grid

19.1 CHAPTER OVERVIEW

We fix an even rational $A = p/q$. In this chapter we prove several results about the graph grid, $G_A = T_A(\mathbf{Z}^2)$. Here

$$T_A \begin{pmatrix} m \\ n \end{pmatrix} = \frac{1}{A+1} \begin{pmatrix} A^2 + A & A+1 \\ -A^2 + 2A + 1 & -2A \end{pmatrix} \begin{pmatrix} m \\ n \end{pmatrix} +$$

$$\begin{pmatrix} \frac{1}{2q} \\ \frac{-1}{2q} + \frac{1}{\omega} + \widehat{\tau} + \omega\theta \end{pmatrix}. \qquad (19.1)$$

T_A is the Canonical Affine Transformation. In §19.2 we prove the Grid Geometry Lemma. In §19.3 we prove the Graph Reconstruction Lemma, a result which describes how the map Ψ_A interacts with G_A.

Computer Tie-In: On the main program, open up the *grid analyzer* window. If you drag the magenta slider back and forth on the top of the *grid analyzer* window, you can see tha grid G_A for varying choices of A. If you click on one of the grid squares in this window, you can see a plot of how the given point of G_A varies with A. The points vary along line segments or hyperbolas.

19.2 THE GRID GEOMETRY LEMMA

Recall that a *distinguished edge* in the graph grid is one connecting distinct points of the form $T(\zeta)$ and $T(\zeta + (i,j))$ for $i, j \in \{-1, 0, 1\}$.

Lemma 19.1 (Grid Geometry) *The following is true.*

1. *No point of the graph grid lies on the boundary of a square.*

2. *Two points of the graph grid cannot lie in the same square.*

3. *A distinguished edge has length less than 2.*

4. *A union of 2 vertically consecutive squares intersects the graph grid.*

5. *A union of 3 horizontally consecutive squares intersects the graph grid.*

maps $\Theta_{A,1}, \Theta_{A,2} : \mathbf{R}^3 \to \mathbf{R}$ as follows.

$$\Theta_{A,1}(x,y,z,A) = x, \qquad \Theta_{A,1}(x,y,z,A) = \frac{y - Ax}{1 + A}. \qquad (19.2)$$

If we replace (x, y, z, A) by any Λ_Y-equivalent vector we get the same answer mod $\mathbf{Z}$. Hence we have a well-defined map from Y_A into $\mathbf{R}/\mathbf{Z}$. Let $\Theta_A : Y_A \to \mathbf{T}^2$ be given by

$$\Theta_A = [(\Theta_{1,A}, \Theta_{2,A})]. \qquad (19.3)$$

Here $[g]$ denotes the image of $g \in \mathbf{R}^2$ in the torus $\mathbf{T}^2 = \mathbf{R}^2/\mathbf{Z}^2$.

Lemma 19.2 (Reconstruction) *For any even rational A and $g \in G_A$, we have* $[g] = \Theta_A \circ \Psi_A(g)$.

Proof: We will give an inductive argument in three steps.

Step 1: A direct calculation shows that the formula holds for $\xi = (0,0)$. In this case both sides of the equation equal

$$\left[\left(\frac{1}{2q}, \frac{q-p}{2q(q+p)} \right) \right].$$

Step 2: We show that if the lemma is true at ξ, it also is true at $\xi \pm dT(0,1)$. Let $P = 2A/(1+A)$. We do the $(+)$ case. The $(-)$ case has the same proof. Let

$$\xi' = \xi + dT(0,1) = \xi + (1, -P).$$

We get $\mu' - \mu = -1$ in Equation 18.4, and this leads to

$$\Psi_A(\zeta') - \Psi_A(\zeta) = (0, 2, 2, 0) \bmod \Lambda_Y.$$

Hence

$$\Theta \circ \Psi(\xi') - \Theta \circ \Psi(\xi) = \Theta(0,2,2,0) = \left(0, \frac{2}{1+A} \right) = [(0, 1 - P)].$$

The last expression is congruent to $\zeta' - \zeta \bmod \mathbf{Z}^2$.

Step 3: We show that if the lemma holds at ξ, it also holds at $\xi \pm dT(1,1)$. We do the $(+)$ case. The $(-)$ case has the same proof. Let

$$\xi' = \xi + dT(1,1) = \xi + (1 + A, 1 - A).$$

We get $\mu' - \mu = 0$ in Equation 18.4, and this leads to

$$\Psi_A(\zeta') - \Psi_A(\zeta) = (1 + A, 1 + A, 1 + A, 0) \bmod \Lambda_Y.$$

Hence

$$\Theta \circ \Psi(\xi') - \Theta \circ \Psi(\xi) = \Theta(1 + A, 1 + A, 1 + A, 0) = [A, 1 - A].$$

The last expression is congruent to $\zeta' - \zeta \bmod \mathbf{Z}^2$. $\square$

Chapter Twenty

The Intertwining Lemma

20.1 CHAPTER OVERVIEW

The purpose of this chapter is to give a proof of the Intertwining Lemma. For convenience, we restate the result here. Let Π be the plaid grid and let $\Pi_A \subset \Pi$ be the set of points which are partnered with points in the graph grid G_A.

Lemma 20.1 (Intertwining) *Let A be an even rational parameter and let $P = 2A/(1 + A)$. Then*

1. $\Phi_A(\Pi_A) \subset Z_P^*$.

2. $\Omega(\Phi_A(c)) = \Psi_A(g)$ *when* $c \in \Pi_A$ *and* $g \in G_A$ *and* $c \to_A g$.

3. Ω *is an injective map from (the interior of) Z^* into Y.*

In §20.2 we list out the formulas for all the maps involved.

In §20.3 we recall the definition of Z^* and prove Statement 3.

In §20.4 we prove Statements 1 and 2 of the Intertwining Lemma for a single point.

In §20.5 we decompose Z^* into two smaller pieces as a prelude to giving the inductive step in our proof.

In §20.6 we prove the following induction step: If the Intertwining Lemma is true for $g \in G_A$ then it is also true for $g + dT_A(0, 1)$.

In §20.7 we explain what needs to be done to finish the proof of the Intertwining Lemma. The result does not quite follow from our single calculation in §20.4 and the inductive step. We need to do more work.

In §20.8 we prove the Intertwining Theorem for points in Π_A corresponding to the points $g_n = (n + 1/2)(1 + A, 1 - A)$ for $n = 0, 1, 2, ...$ which (as we check) all belong to G_A. This result combines with our induction step to finish the proof, as explained in §20.7.

Remark: I'd like to apologize to the reader for the painful proof of the Intertwining Lemma. This is a result that I discovered on the computer, and unfortunately I could not find a short proof.

20.2 A RESUME OF TRANSFORMATIONS

Here are the basic maps involved in the Intertwining Theorem. Here are the basic quantities.

$$A = \frac{p}{q}, \qquad \omega = p + q, \qquad P = \frac{2p}{\omega} = \frac{2A}{1 + A}. \qquad (20.1)$$

We also let $\hat{\tau} \in (0, \omega)$ be the solution to $2p\hat{\tau} \equiv 1 \mod \omega$. Finally, we set $\theta = 0$ or $\theta = 1$ according as to whether $\hat{\tau}$ is odd or even.

Here is the formula for $\Phi_A : \Pi \to X$.

$$\Phi_A(x, y) = x(2P, 2P, 2P, 0) + y(2, 0, 2P, 0) + (0, 0, 0, P). \qquad (20.2)$$

To get a well-defined point in the space X_P we mod out by the lattice Λ_X generated by the translations

$$(4, 2P, 2P, 0), \qquad (0, 2, 0, 0), \qquad (0, 0, 2, 0). \qquad (20.3)$$

Technically Λ_X is a group of affine automorphisms of $\widehat{X} = \mathbf{R}^3 \times [0, 1]$, but when we restrict to the slice $\widehat{X}_P = \mathbf{R}^3 \times \{P\}$ the action is by translations.

Here is the formula for $\Psi_A : G_A \to Y$.

$$\Psi_A(x, y) = (x + \mu, x - \mu, x - \mu, A)$$

$$\mu(x, y) = 1 + \frac{x(A - 1) + y(A + 1)}{A + 1}. \qquad (20.4)$$

To get a well-defined point in the space Y_A we quotient out by the lattice Λ_Y generated by the translations

$$(2, -2, -2, 0), \qquad (0, 1 + A, 1 - A, 0), \qquad (0, 0, 1 + A, 0). \qquad (20.5)$$

The same remarks about affine transformations versus translations apply to Λ_Y.

Here is the formula for $\Omega : Z^* \to Y$. We take a representative of our points to be of the form $(x, y, z, P) \in \widehat{Z}_\pm$, then define

$$\widehat{\Omega}(x, y, z, P) = \frac{1}{2 - P}\Big((x - y, -y, z, P) \pm (-2 + P, 1, 1 - P, 0)\Big), \qquad (20.6)$$

then quotient out by Λ_Y to get a well-defined point in Y.

Here is the formula for the Canonical Affine Transformation T_A:

$$T_A \begin{pmatrix} m \\ n \end{pmatrix} = \frac{1}{A + 1} \begin{pmatrix} A^2 + A & A + 1 \\ -A^2 + 2A + 1 & -2A \end{pmatrix} \begin{pmatrix} m \\ n \end{pmatrix} +$$

$$\begin{pmatrix} \frac{1}{2q} \\ \frac{-1}{2q} + \frac{1}{\omega} + \hat{\tau} + \omega\theta \end{pmatrix}. \qquad (20.7)$$

20.3 INJECTIVITY OF THE MAP

Here we prove that Ω maps the interior of Z^* into Y. We first recall the definition of Z^*. We have $Z^* = Z^*_+ \cup Z^*_-$ with $Z^*_- = I(Z^*_+)$. Here

$$I(x, y, z, P) = (-x, -y, -z, P).$$

Recall also that

$$Z^*_+ = \bigcup_{P \in [0,1]} Z^*_{P,+}, \qquad Z^*_P = H^*_{P,+} \times \mathbf{R}. \qquad (20.8)$$

Here $H^*_{P,+}$ is the parallelogram with vertices

$$(P - 1, P - 1), \ (1 - P, -1), \ (3 - P, 1), \ (P + 1, P + 1).$$

We introduce the simpler map

$$\Omega^*(x, y) = \frac{1}{2 - P}(x - y, -y). \qquad (20.9)$$

This is just the first two coordinates of the linear part of Ω. The set $\Omega(Z^*_{P,\pm}) \subset \mathbf{R}^3 \times \{A\}$ is just a translate of $\Omega^*(H_{P,\pm}) \times \mathbf{R}$. To prove our injectivity result, we just have to prove that

$$\Omega^*(H_{P,+}) \cup \Omega^*(H_{P,-})$$

is contained in a fundamental domain for the action of the planar lattice generated by $(2, -2)$ and $(0, 1 + A)$. These vectors are obtained from the vectors in Equation 20.5 by dropping the last two coordinates.

To simplify our calculation further, we observe that $\Omega(Z^*_\pm) \subset \Omega(Z_\pm)$, and these latter sets have disjoint interiors. Hence, it suffices to show that $\Omega^*(H_{P,+})$ is contained in a fundamental domain for our planar lattice. We will show more precisely that $\Omega^*(H_{P,+})$ is a fundamental domain for the lattice λ_A generated by $(1, -1)$ and $(0, 1 + A)$. We compute that the vertices of $\Omega^*(H_{P,+})$ are

$$\left(0, \frac{1 - A}{2}\right), \ \left(1, \frac{1 + A}{2}\right), \ \left(1, \frac{-1 - A}{2}\right), \ \left(0, \frac{-1 - 3A}{2}\right). \qquad (20.10)$$

One can see that these vertices are identified in pairs under the basis vectors of λ_A. Hence, $\Omega^*(H_{P,+})$ is a fundamental domain for the action of λ_A. This completes the proof.

20.4 CALCULATING A SINGLE POINT

Here we prove the Intertwining Lemma for the pair of points

$$c = \left(\frac{1}{2}, \frac{1}{2}\right), \qquad g = \left(\frac{1 + A}{2}, \frac{1 - A}{2}\right). \qquad (20.11)$$

Lemma 20.2 $g \in G_A$.

Proof: Let $\widehat{\tau}$ be as in §1.2 and let $\theta = 0$ or $\theta = 1$ according as to whether $\widehat{\tau}$ is odd or even. Let

$$\zeta = \left(\frac{1}{2}, \frac{1}{2}\right) + \left(\frac{-1 - 2q\widehat{\tau}}{2\omega}, \frac{-1 + 2p\widehat{\tau}}{2\omega}\right) + \theta(-q, p).$$

The second summand is a vector having half-integer coordinates. Hence, $\zeta \in \mathbf{Z}^2$. We compute that $T_A(\zeta) = g$. Hence $g \in G_A$. □

We evidently have $c \to_A g$ for all A. We compute

$$\Phi_A(c) = (P + 1, P, 2P, P) \in Z_+^*.$$

Next, we compute

$$\widehat{\Omega} \circ \Phi_A(c) - \Psi_A(g) =$$

$$\widehat{\Omega}(P + 1, P, 2P, 0) - (1 + A, 1 + A, 1 + A, A) + (1, -1, -1, 0) =$$

$$(-2, 1 - A, 1 + A, 0) = (-2, 2, 2, 0) + (0, -1 - A, -1 + A, 0).$$

This last vector is 0 mod Λ_Y. So, Statements 1 and 2 hold for the pair (c, g) for all even rational parameters A.

20.5 DISSECTING THE SET

Recall that from Equation 18.5 that Z_+^* is defined in terms of the polygon $H_{+,P}^*$. Let $H_{+,P}^{\mathrm{lo}}$ denote the parallelogram with vertices

$$(3 - 3P, 1 - 2P), \quad (3 - P, 1), \quad (1 + P, 1 + P), \quad (1 - P, 1 - P). \quad (20.12)$$

Let $H_{+,P}^{\mathrm{hi}}$ denote the parallelogram with vertices

$$(-1 + P, -1 + P), \quad (1 - P, -1), \quad (3 - 3P, 1 - 2P), \quad (1 - P, 1 - P). \quad (20.13)$$

We have

$$H_{+,P}^* = H_{+,P}^{\mathrm{lo}} \cup H_{+,P}^{\mathrm{hi}}. \quad (20.14)$$

Correspondingly, we have a decomposition of prisms:

$$Z_{+,P}^* = Z_{+,P}^{\mathrm{lo}} \cup Z_{+,P}^{\mathrm{hi}}. \quad (20.15)$$

Figure 20.1: $H_{P,+}^{\mathrm{lo}}$ (dark) and $H_{P,+}^{\mathrm{hi}}$ (light) for $P = 1/2$.

We make a similar dissection or Z_-^*, using the sets

$$Z_-^{\text{hi}} = Z_+^{\text{hi}} - (1, -1, 1, 0), \qquad Z_-^{\text{lo}} = Z_+^{\text{lo}} - (1, -1, 1, 0). \qquad (20.16)$$

Computer Tie-In: On the main program, open up the *Plaid PET* window and the *Plaid PET* control panel. Turn off all the features except *hit set hi* and *hit set lo*. On the *plaid PET slicer*, set the *slice* to Z and the *parameter* to *free*. Now drag the mouse around the *plaid PET slicer*. This will let you see pictures like Figure 20.1 for any parameter.

Now we come to geometric lemma about these sets. Let $g \in G_A$ be some point in the graph grid. We say that G_A^{hi} consists of those $(a, b) \in G_A$ such that $[y] > P$ and G_A^{lo} consists of those $(a, b) \in G_A$ such that $[y] < P$. Here $[y]$ is the image of y in the fundamental domain $[0, 1)$ for $\mathbf{R}/\mathbf{Z}$. The case $[y] = 0$ never occurs in our setting, thanks to the Grid Geometry Lemma.

Lemma 20.3 *Let $c \in \Pi_A$ and $g \in G_A$ be such that $c \to_A g$. Suppose that the Intertwining Lemma is true for (c, g). Then, modulo Λ_X', the following is true.*

- *If $g \in G_A^{\text{hi}}$ then $\Phi_A(c) \subset Z_P^{\text{hi}}$.*

- *If $g \in G_A^{\text{lo}}$ then $\Phi_A(c) \subset Z_P^{\text{lo}}$.*

Proof: Let $g = (a, b)$ and $\Phi_A(c) = (x, y, z, P)$. We consider the case when $(x, y, z, P) \in Z_{P,+}^*$. The case when $(x, y, z, P) \in Z_{P,-}^*$ has the same kind of proof. Let $(x^*, y^*, z^*, A) = \Omega(x, y, z, P)$. Since the Intertwining Lemma holds for (c, g) we have $(x^*, y^*, z^*, A) = \Psi_A(g)$. By the Graph Reconstruction Lemma,

$$[a] = [x^*], \qquad [b] = \left[\frac{y^* - Ax^*}{1 + A}\right].$$

Plugging this into the formula for $\widehat{\Omega}$ on $\widehat{Z}_+$, we get

$$[a] = \left[\frac{x - y}{2 - P}\right], \qquad [b] = \left[\frac{-2 - P - P^2 + Px + 2y - 2Py}{2P - 4}\right]. \qquad (20.17)$$

Equation 20.17 does not depend on the Z-coordinate, so we intrepret Equation 20.17 as a map $\mho$ from the (x, y) coordinates on $H_{P,+}^*$ to the (a, b) coordinates on $[0, 1]^2$. We compute that $\mho$ maps the vertices of $H_{P,+}^{\text{lo}}$ to

$$(1, P), \qquad (1, 0), \qquad (0, 0), \qquad (0, P).$$

We compute that $\mho$ maps the vertices of $H_{P,+}^{\text{hi}}$ to

$$(0, 1), \qquad (1, 1), \qquad (1, P), \qquad (0, P).$$

These calculations, together with the affine nature of the formula for $\mho$, shows that $\mho$ is an affine diffeomorphism from $H_{P,+}^*$ to $[0, 1]^2$, and

$$H_{P,+}^{\text{hi}} = \mho^{-1}([0, 1] \times [P, 1]), \qquad H_{P,+}^{\text{lo}} = \mho^{-1}([0, 1] \times [0, P]). \qquad (20.18)$$

This is equivalent to what we wanted to prove. $\square$

20.6 THE INDUCTION STEP

Let $g' = g + dT_A(0,1)$. Let $c, c' \in \Pi_A$ be such that $c \to_A g$ and $c' \to_A g'$.

Lemma 20.4 *If Statement 1 of the Intertwining Lemma is true for (c, g) then it is also true for (c', g').*

Proof: We will treat the case when $\Phi_A(c) \in Z^*_{P,+}$. The other case is similar. As in §19.2 we have

$$dT_A(0,1) = (1, -P). \tag{20.19}$$

From this equation, we see that either $c' - c = (1, 0)$ or $c' - c = (1, -1)$ depending on whether $g \in G^{\mathrm{hi}}_A$ or $g \in G^{\mathrm{lo}}_A$. The Grid Geometry Lemma rules out the boundary case.

We introduce the new sets

$$(H^{\mathrm{hi}}_{P,+})' = H^{\mathrm{hi}}_{P,+} + (2P, 2P), \qquad (H^{\mathrm{lo}}_{P,+})' = H^{\mathrm{lo}}_{P,+} + (2P - 2, 2P - 2). \tag{20.20}$$

Likewise we define the products $(Z^{\mathrm{hi}}_{P,+})'$ and $(Z^{\mathrm{lo}}_{P,+})'$. We have

$$(Z^{\mathrm{lo}})' \cup (Z^{\mathrm{hi}})' = Z^{\mathrm{lo}} \cup Z^{\mathrm{hi}}. \tag{20.21}$$

If $g \in G^{\mathrm{hi}}_A$ then

$$\Phi_A(c') \in Z^{\mathrm{hi}}_{P,+} + (2P, 2P, 2P, 0) = (Z^{\mathrm{hi}}_{P,+})' \bmod \Lambda_X.$$

If $g \in G^{\mathrm{lo}}_A$ then

$$\Phi_A(c') \in Z^{\mathrm{lo}}_{P,+} + (2P - 2, 2P, 0) = Z^{\mathrm{lo}}_{P,+} + (2P - 2, 2P - 2, 0) \bmod \Lambda_X.$$

So, in all cases, $\Phi_A(c') \in Z^*_{P,+}$. □

Lemma 20.5 *If Statement 2 of the Intertwining Lemma is true for (c, g) then it is also true for (c', g').*

Proof: As in Step 2 of the proof of the Graph Reconstruction Lemma in §19.3, we have

$$\Psi_A(g') - \Psi_A(g) = (0, 2, 2, 0) \bmod \Lambda_Y. \tag{20.22}$$

Let

$$\phi = \Phi_A(c), \qquad \phi' = \Phi_A(c'). \tag{20.23}$$

To finish the proof, we just have to show that

$$\Omega(\phi') - \Omega(\phi) - (0, 2, 2, 0)$$

is equivalent to 0 mod Λ_Y. There are two cases. Before we consider these cases, we make one definition. Call ϕ and ϕ' *coherent* if they both lie in the same component of $Z^*_+ \cup Z^*_-$.

Case 1: Suppose that $g \in G_A^{\mathrm{hi}}$. The analysis in the previous lemma shows that we can take coherent representatives such that

$$\phi' - \phi = (2P, 2P, 2P + 2\beta, 0) \tag{20.24}$$

and either $\phi, \phi' \in Z_+^*$ or $\phi, \phi' \in Z_-^*$. In either case, we use the same branch of Ω when we compute $\Omega(\phi)$ and $\Omega(\phi')$. Hence $\Omega(\phi) - \Omega(\phi') = d\Omega(\phi - \phi')$. Here $d\Omega$ is the linear part of Ω:

$$d\Omega(x, y, z, P) = (x - y, -y, z, P).$$

We have

$$\Omega(\phi') - \Omega(\phi) - (0, 2, 2, 0) =$$

$$d\Omega(\phi' - \phi) - (0, 2, 2, 0) =$$

$$d\Omega(2P, 2P, 2P + 2\beta, 0) - (0, 2, 2, 0) =$$

$$\frac{1}{2 - P}\Big(0, -2P, 2P + 2\beta, 0\Big) - (0, 2, 2, 0) =$$

$$(0, -2A - 2, 2A - 2 + \beta(1 + A), 0) =$$

$$-2(0, 1 + A, 1 - A, 0) + \beta(0, 0, 1 + A, 0).$$

This vector is equivalent to $0 \bmod \Lambda_Y$.

Case 2: Suppose that $g \in G_A^{\mathrm{lo}}$. The analysis in the previous lemma shows that we can take coherent representatives such that

$$\phi' - \phi = (2P - 2, 2P - 2, 2\beta, 0). \tag{20.25}$$

We have

$$\Omega(\phi') - \Omega(\phi) - (0, 2, 2, 0) =$$

$$d\Omega(\phi' - \phi) - (0, 2, 2, 0) =$$

$$d\Omega(2P - 2, 2P - 2, 2\beta, 0) - (0, 2, 2, 0) =$$

$$\frac{1}{2 - P}\Big(0, -2P + 2, 2\beta, 0\Big) - (0, 2, 2, 0) =$$

$$(0, -1 - A, -2 + \beta(1 + A), 0) =$$

$$-(0, 1 + A, 1 - A, 0) + (\beta - 1)(0, 0, 1 + A, 0).$$

This vector is equivalent to $0 \bmod \Lambda_Y$. $\square$

20.7 DISCUSSION

It would be nice if we had an inductive step that would allow us to conclude the Intertwining Lemma from the calculation in §20.4. However, the induction step from the previous section just tells us about the points $g+dT_A(0,m)$ for all $m = 1, 2, 3, ...$ This is a fairly thin subset of G_A.

One approach to finishing the proof would be to prove that if the Intertwining Lemma holds for some point $g' \in G_A$ then it also holds for $g'+dT_A(1,0)$. We could certainly do this, though the proof might be just as painful as what we end up doing below. The problem is that the vector $dT_A(1,0)$ is much more difficult to work with than the vector $dT_A(0,1)$. We would end up getting a more complicated decomposition of Z^* and giving a more elaborate version of the argument above. The other vectors $dT_A(i,j)$ for $|i|, |j| \leq 1$ present similar problems. The vector $dT_A(q,q)$ works very well, but it doesn't give us enough information.

Instead, we will finish the proof by showing directly that the Intertwining Lemma holds for the infinite set of points

$$g_n = \left(n + \frac{1}{2}\right)(1 + A, 1 - A), \qquad n = 0, 1, 2, ... \qquad (20.26)$$

With this result we can finish the proof of the Intertwining Lemma. The set S of points in the graph grid G for which we have proved the Intertwining Lemma is the intersection of G with a nontrivial cone. S contains every point of G up to the lattice L_A generated by the vectors $(\omega^2, 0)$ and $(0, \omega)$. This lattice is a symmetry of the entire construction. So, since the Intertwining Lemma holds on a subset of G which contains every L_A-orbit, it holds on all of G. By symmetry, it holds everywhere.

20.8 THE DIAGONAL CASE

We fix an integer $n > 0$. In this section we prove the Intertwining Lemma for $g_n = (n + 1/2)(1 + A, 1 - A)$. Since $dT_A(1,1) = (1 + A, 1 - A)$ and $g_0 \in G_A$, we see that $g_n \in G_A$.

The point g_n sweeps through different unit integer squares as a function of A, moving in a straight line of slope -1. We divide up the parameter interval so as to capture this structure. We define the (parameter) sub-intervals

$$R_{n,k} = \left(\frac{2k}{2n+1}, \frac{2k+1}{2n+1}\right), \qquad k = 0, ..., (n-1). \qquad (20.27)$$

$$L_{n,k} = \left(\frac{2k-1}{2n+1}, \frac{2k}{2n+1}\right), \qquad k = 1, ..., (n-1). \qquad (20.28)$$

We ignore the endpoints of these intervals; the boundary cases, when relevant, follow from continuity. The endpoints correspond to the parameters where one or the other coordinate of g_n is an integer.

When $A \in L_{n,k}$ (respectively $A \in R_{n,k}$) we have $c_{n,k} \to_A g_n$, where

$$c_{n,k} = \left(n + \frac{1}{2}, n + \frac{1}{2}\right) + (k, -k), \tag{20.29}$$

and g_n lies in the left (respectively right) half of the square $\square_c$ where $c = c_{n,k}$.

The Boundary Trick: Let $I = (A_0, A_1)$ be one of the intervals of interest. We will sometimes use the following trick to bound certain numerical quantities that depend on $A \in I$. When the quantity is monotone, we get the bounds by evaluating the expression on the boundary values $A = A_0$ and $A = A_1$. We call this method the *boundary trick*.

Lemma 20.6 *Statement 1 of the Intertwining Theorem holds for $c_{n,k}$ for all parameters. That is, $\Phi_A(c_{n,k}) \subset Z^*$ for all even rational A.*

Proof: We fix A and set $P = 2A/(1 + A)$. We work in a fixed P slice so that we may interpret Λ_X as the group of translations. Let Λ_X' denote the lattice of vectors generated by $(2, P, P, 0)$, $(0, 2, 0, 0)$ and $(0, 0, 2, 0)$. Here Λ_X' contains Λ_X with index 2. We will compute $\Phi_A(c_{n,k})$ modulo Λ_X' and then deduce the location of $\Phi_A(c_{n,k})$ mod Λ_X at the end.

We have

$$\Phi_A(c_{n,k}) \equiv (P(2n+2k+1)+(2n-2k+1), P(2n+2k+1), 2P(2n+1)). \tag{20.30}$$

The symbol $\equiv$ means that we still need to reduce mod Λ_X' to get a vector in the fundamental domain.

The boundary trick tells us that the first coordinate in Equation 20.30 lies in

$$\left(1 + 2k + 2n - \frac{1 + 2n}{k + n}, 1 + 2k + 2n\right). \tag{20.31}$$

So, we subtract off the lattice vector $(n + k)(2, P, P)$. This gives

$$\Phi_A(c_{n,k}) \equiv (P(2n + 2k + 1) - 4k + 1, P(1 + k + n), 2P(2 - k + 3n)). \tag{20.32}$$

The boundary trick tells us that the second coordinate in Equation 20.32 lies in the interval $(-1, 1) + 2k$. Subtracting $(0, 2k, 0)$ from Equation 20.32, we get a point in the fundamental domain:

$$\Phi_A(c_{n,k}) = (P(2n + 2k + 1) - 4k + 1, P(1 + k + n) - 2k, P(2 - k + 3n) + 2\beta). \tag{20.33}$$

Here β is some integer whose value we don't care about.

The boundary trick, applied to the difference of the first two coordinates, shows that the first coordinate in Equation 20.33 is larger than the second coordinate.

Suppose first that $A \in L_{n,k}$. Consider the triangle Δ_L with vertices

$$(-1 + P, -1 + P), \qquad (1, -1 + P), \qquad (1, 1) \tag{20.34}$$

is the convex hull of 3 of the vertices and is contained in the polygon $H_{P,+}$ defined in Equation 18.4. The left side of Figure 20.2 shows Δ_L. The square drawn on the left side of Figure 20.2 is $[-1, 1]^2$, a slice of the fundamental domain for Λ_X'. (The figure is drawn for the the parameter $P = 1/2$.)

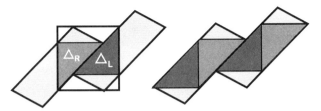

Figure 20.2: (Z, P) The triangles Δ_L and Δ_R.

Let ξ denote the point in the plane obtained by taking the first two coordinates of the point in Equation 20.33. That is:

$$\xi = (P(2n + 2k + 1) - 4k + 1, P(1 + k + n) - 2k). \qquad (20.35)$$

We show that $\xi \in \Delta_L$ for all $A \in L_{n,k}$. One of the sides of Δ_L is the line $x = 1$, and certainly ξ does not cross this line because this line contains a side of the fundamental domain for Λ'_X. Another side of Δ_L is the line $y = -1 + P$. The boundary trick shows that ξ stays above this line. Finally, the other side of Δ_L is the line $y = x$. We already know that the first coordinate of ξ is greater than the second, and this is the equivalent to the statement that ξ lies to the right of this line. These three conditions together imply that $\xi \in \Delta_L$. When we lift to the double cover, we see that ξ lies in

$$\Delta_L \cup \left(\Delta_L - (2, P, P) \right).$$

These are the darkly shaded triangles on the right side of Figure 20.2. Both triangles belong to H_P^*. Hence $\Phi_A(c_{n,k}) \in Z_P^*$.

Now suppose $A \in R_{n,k}$. This time we use the triangle $\Delta_R = -\Delta_L$, obtained by negating all the coordinates of the vertices of Δ. This time we want to show that the point

$$(P(2k + 2n + 1) - 4k - 1, P(k + n) - 2k) \qquad (20.36)$$

lies in Δ_R. The same arguments as above show that this point lies to the right of the line $x = -1$, below the line $y = 1 - P$, and to the left of the line $y = x$. This does the job for us. This time, when we lift to the double cover, $\Phi_A(c_{n,k})$ lies in the union of the two lighter shaded triangles shown on the right side of Figure 20.2. $\square$

Lemma 20.7 *Statement 2 of the Intertwining Lemma holds for $c_{n,k}$ and g_n.*

Proof: We treat the case when $A \in L_{n,k}$ in detail and then briefly describe the small differences for the other case. We first consider the case when $n + k$ is even. In this case, the reduction from Equation 20.30 to Equation 20.32

still works because we can subtract off the vector $(n + k)/2 \times (4, 2P, 2P)$. So, mod Λ_X, we still have

$$\Phi_A(c_{n,k}) = (P(2n + 2k + 1) - 4k + 1, P(1 + k + n) - 2k, P(2 - k + 3n) + 2\beta). \tag{20.37}$$

At the same time, we have

$$\Psi_A(g_n) \equiv \left(n + \frac{1}{2}\right)(1 + A, 1 + A, 1 + A) + (1, -1, -1) \quad \text{mod } \Lambda_Y. \tag{20.38}$$

Referring to Equation 18.4, the key fact is that $\mu(g_n) = 1$ for all n. A direct calculation shows that

$$\Psi_A(g_n) - \widehat{\Omega}(\Phi_A(c_{n,k})) =$$

$$(2 + k + n)(1, -1, -1) +$$

$$(1 + 2n)(0, 1 + A, 1 - A) +$$

$$(1 + k + \beta)(0, 0, 1 + A).$$

This expresses the difference as an integer combination of basis vectors for the lattice Λ_Y as it acts on the A slice. Hence $\Omega \circ \phi)A(c_{n,k}) = \Psi_A(g_n)$.

When $n + k$ is odd, the expression for $\Phi_A(c_{n-k})$ is the same, except that we need to add the vector $(1, -1, -1)$. In this case we get the same linear combination except that the first coefficent is $1 + k + n$. This is again an even number, so we get the same final result.

When $A \in R_{n,k}$ we start with the case when $n + k$ is odd. In this case, the reduction from Equation 20.31 to Equation 20.32 involves subtracting $(n + k + 1)(2, P, P)$. We can still do this mod Λ_X because $(n + k + 1)$ is even. the rest of the reduction is the same, and we get to the equation

$$\Phi_A(c_{n,k}) = (P(2k + 2n + 1) - 4k - 1, P(k + n) - 2k, P(1 - k + 3n) + \beta). \tag{20.39}$$

This time we use the branch of Ω defined in Z_-. A similar computation shows that $\Omega \circ \phi)A(c_{n,k}) = \Psi_A(g_n)$. The odd-to-even modification for the case $A \in R_{n,k}$ is the same as the even-to-odd modification for the case $A \in L_{n,k}$. $\square$

Part 5. The Distribution of Orbits

Chapter Twenty-One

Existence of Infinite Orbits

21.1 CHAPTER OVERVIEW

This chapter begins Part 5 of the monograph. This part is devoted mostly to the study of the distribution of the plaid polygons: their size and number depending on the parameter.

In this chapter we will give a light proof of the following result.

Theorem 21.1 *Let $P \in (0,1)$ be irrational and let X_P be the P slice of the plaid PET X. Then X_P contains a point with a well-defined and infinite orbit.*

Theorem 21.1 is a weaker version of Theorem 0.7, but it doesn't take as much effort to prove. One might expect that Theorem 21.1 is true because Theorem 3.1 tells us that the slice $X_{p/q}$ has an orbit with period greater than $p + q$. The difficulty with simply taking the limit of these orbits is that the limit might not have a well-defined orbit. We will see how to get around this problem.

In §21.2 we give a criterion for a point in F_X to have a well-defined orbit.

In §21.3 we revisit the pixelated spacetime diagrams of capacity 2, and use them to construct a large supply of plaid polygons having large diameter. One could view the analysis here as a generalization of the analysis from Theorem 3.1.

Our construction in §21.3 works one parameter at a time. In §21.4 we take the limit of our construction relative to a sequence of even rational parameters converging to our irrational parameter. This limiting argument completes the proof.

In §21.5 we explain how to associate a plaid path to an infinite orbit. Recall that a path in the plane is *fat* if it does not lie in an infinite strip. One shortcoming of our proof of Theorem 21.1 is that the path associated to our infinite orbit might not be fat. We might want the associated path to be fat so that its projection into the X-axis, which models some special outer billiards orbit, is unbounded.

In §21.6 we give a quick alternate proof of Theorem 21.1, based on results from [**S1**], in which we can draw the conclusion that the associated plaid path is fat. This statement is closer to Theorem 0.7 from the introduction. Of course, our proof has the disadvantage of relying on all the work done in [**S1**].

In subsequent chapters we will give an independent proof of Theorem 0.7.

21.2 DEFINEDNESS CRITERION

Here we give a criterion for a point of the plaid PET X to have a well-defined orbit. Define

$$\boldsymbol{Q}[P] = \{r_1 + r_2 P \mid r_1, r_2 \in \boldsymbol{Q}\}. \tag{21.1}$$

Lemma 21.2 (Definedness Criterion) *Suppose $V = (x, y, z, P)$ satisfies one of two properties.*

1. $x \in \boldsymbol{Q}[P]$ *and* $y, z \notin \boldsymbol{Q}[P]$.

2. $x \notin \boldsymbol{Q}[P]$ *and* $y, z \in \boldsymbol{Q}[P]$.

Proof: Note first that our criterion is independent of the representative we take mod the group Λ_X, because this group acts by integral affine transformations. For this proof we ignore the last coordinate and think of the slice X_P as a 3-torus. We think of X_P as a fiber bundle over the X-fiber. The fiber over the point x intersects the walls of our partition in rectangles of the form $y = y_0$ and $z = z_0$, where y_0 and z_0 are various numbers in the translated set $\boldsymbol{Q}[P] + x$. This is a consequence of the fact that the 4-dimensional polytopes of our partition are integral.

Let $V_0 = V$. If V_0 satisfies either of our conditions, then V_0 does not lie in a partition wall. Hence F_X is defined on V_0. All the continuous branches of F_X, when acting on X_P, are translations by vectors with coordinates in $\boldsymbol{Q}[P]$. Hence $V_1 = F_X(V_0)$ satisfies the same criterion as V_0. Hence F_X is defined on V_1. And so on. The same argument works for the inverse map. $\square$

21.3 SPACETIME DIAGRAMS REVISITED

We fix a parameter $A = p/q$ and consider a pixelated horizontal spacetime diagram of capacity 2 associated to A. Specifically, we take the one corresponding to the horizontal line $y = \hat{\tau}$. Figure 21.1 shows what we mean for the parameter $p/q = 2/5$. Actually, there is a second half to the spacetime diagram, but we will follow the same custom as in §5 and just show half the picture.

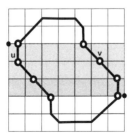

Figure 21.1: The good partner points in the capacity 2 slice.

The shaded band in Figure 21.1 corresponds to points on the loop where both of the corresponding particles have type Q. This band has width $\omega = p + q$ and height $q - p$ in general. Each marked point in Figure 21.1 corresponds to the midpoint of a horizontal unit integer segment containing exactly one light point of type Q. Some plaid polygon goes through such a point. For instance, let π_u be the plaid polygon containing the point corresponding to the point u in the picture.

Two marked points on the same horizontal line correspond to points lying on the same plaid polygon, because these two points lie in the same block and on the same horizontal line of capacity 2. Only one such plaid polygon intersects such a line. In short, we are saying that $\pi_u = \pi_v$, and likewise for the other pairs of points. From this we see that π_u has diameter at least $\omega/2$, because u and v correspond to points that are at least $\omega/2$ apart. The same goes for the other plaid polygons associated to these marked points.

More generally, let C_A denote the set of marked vertices like this which lie on the left side of the loop. For instance, $C_{2/5}$ contains 4 points, one of which is the point u. Let Λ_A denote the set of plaid polygons associated to points in C_A. Thus, $\Lambda_{2/5}$ consists of 4 plaid polygons. In general, Λ_A consists of $q - p + 1$ plaid polygons, each having width at least $\omega/2$.

Now we study the image

$$S_A = \Phi_A(C_A). \tag{21.2}$$

Here are some general properties of S_A:

- By Statement 4 of Lemma 9.1, the Segment Lemma, the set S_A lies in a certain segment Σ_A parallel to the vector $(1, 0, 0)$.

- Inspecting the proof of Lemma 9.1, we see that the points of S_A are evenly spaced on Σ_A.

- The length of Σ_A is at least $(q - p)/(p + q) = (1 - A)/A$.

Now we compute the endpoint of S_A corresponding to the point $(1/2, \widehat{\tau})$. This point is the topmost point of C_A. To make our calculation simpler, we compute mod the larger lattice Λ'_X generated by the vectors $(2, P, P, 0)$ and $(0, 2, 0, 0)$ and $(0, 0, 2, 0)$. We have

$$\Phi_A(1/2, \widehat{\tau}) = (P, P, P, P) + (2\widehat{\tau}, 0, 2P\widehat{\tau}, 0) \equiv (P, P, P, P) + (0, -P\widehat{\tau}, P\widehat{\tau}, 0).$$

Note that there is some odd integer k such that

$$P\widehat{\tau} = \frac{2p\widehat{\tau}}{\omega} = \frac{k\omega + 1}{\omega} = k + \frac{1}{\omega}.$$

The upshot is that one endpoint of S_A is within $1/\omega$ from being the point $(P, P - 1, P + 1, P)$.

21.4 TAKING A LIMIT

We now give a criterion for a point in x having an infinite F_X-orbit. In the next lemma and its proof, we set $\Phi_n = \Phi_{p_n/q_n}$.

Lemma 21.3 *Let $A \in (0,1)$ be irrational and let $\{p_n/q_n\}$ be a sequence of even rationals converging to A. Let π_n be a plaid polygon associated to p_n/q_n and let c_n be a vertex of π_n. Suppose that*

1. *$\zeta = \lim_{n \to \infty} \Phi_n(c_n)$ exists in X and has a well-defined F_X-orbit.*

2. *The length of π_n converges to ∞ as $n \to \infty$.*

Then ζ has an infinite orbit.

Proof: Suppose that ζ has a finite F_X-orbit. Then there is a convex polytope τ such that ζ lies in the interior of τ and all points of τ are periodic with the same period. For n sufficiently large, the point $\zeta_n = \Phi_n(c_n)$ lies in the interior of τ. But then, by the Plaid Master Picture Theorem, the length of π_n equals the period of points in τ. This is a contradiction. $\square$

Now we take a limit of the construction made in the previous section. Let $\{p_n/q_n\}$ be a sequence of even rationals converging to A. We set $C_n = C_{A_n}$, etc. The number of points in C_n tends to ∞ with n. As $n \to \infty$, the geometric limit of the sequence $\{S_n\}$ is a line segment Σ_A parallel to $(1,0,0,0)$ and having one endpoint $(P, P-1, P+1, P)$.

But this means that we can choose a sequence $\{c_n\}$ with $c_n \in C_n$ such that $\lim \Phi_n(c_n)$ is a point ζ of the form $(x, P-1, P+1, P)$ with $x \notin \mathbf{Q}[P]$. By our definedness criterion, ζ has a well-defined orbit. Since the diameter of the plaid polygon π_n through c_n is at least $\omega_n/2$, a quantity which tends to ∞ with n, Lemma 21.3 says that ζ has an infinite orbit. This completes the proof.

21.5 ASSOCIATED PATHS

Recall that our plaid PET F_X is defined in terms of a pair of partitions of X. Each polytope τ in the first partition has a translation vector V_τ such that $F_X(\zeta) = \zeta + V_\tau$ for all $\zeta \in \tau$. The polytope τ also has a label $L_\tau \in \{NE, SW, ...\}$ with the following property. If A is an even rational parameter and $c \in \Pi$, the plaid grid, then $\Phi_A(c) \in \tau$ implies that the tile in the plaid model connects the sides of the unit integer square $\square_c$ centered at c according to the label L_τ. For instance, if $L = NW$ then the plaid polygon in $\square_c$ connects the north side to the west side.

Let O be some orbit of F_X. We write $O = \{x_i\}$. For each i there is a polytope τ_i in the first partition of X, and a corresponding label L_i. We build a path in the plane as follows. Modulo translation, there is a unique way to choose a sequence $\{\square_j\}$ of unit integer squares so that

- $\square_i$ is decorated with the directed edge corresponding to the label L_i.

- $\square_i$ and $\square_{i+1}$ share the edge given by the second label of L_i. For instance, if L_i is NW then $\square_{i+1}$ attaches to $\square_i$ across the west edge of L_i.

The path made from the union of connectors in these tiles is the path associated to O.

We can think of the associated path as a geometric limit. There is some A such that $O \subset X_A$, the A slice of X. Let $\{A_n\}$ be a sequence of even rational parameters which converges to A. Let $\Phi_n = \Phi_{A_n}$ be the map from the Plaid Master Picture Theorem. Let $X_n = X_{A_n}$. Let Π be the plaid grid. Thanks to Lemma 8.1, the image $\Phi_n(\Pi)$ becomes dense in X_n as $n \to \infty$. So, we can choose a sequence $\{c_n\} \subset \Pi$ such that $\Phi_n(c_n) \to x_0$, the point whose orbit we care about. Let π_n be the plaid polygon through the point c_n.

Lemma 21.4 *The path associated to the orbit $\{x_n\}$ coincides with the geometric limit of the sequence $\{\pi'_n\}$. Here π'_n is a suitable translate of π_n.*

Proof: The point x_0 is contained in the nested intersection

$$\sigma_k = \bigcap_{i=-k}^{k} F_X^i(\tau_i). \tag{21.3}$$

Since x_0 is part of a well-defined orbit, x_0 lies in the interior of σ_k for all k. For any fixed k, there is some N such that $n > N$ implies that $\Phi_n(c_n) \in \sigma_k$. But then the first k steps of π_n in either direction have the same combinatorial structure as the first k steps of π in either direction. For instance, if the 3rd step of π_n in the forward direction involves a NW connector, so does the 3rd step of π in the forward direction. If we translate π_n so that c_n coincides with the initial square $\square_0$ of π, then π and π'_n agree for their first k steps in either direction. Letting $n \to \infty$ we have $k \to \infty$ as well, and we get the desired convergence. $\square$

As we discussed above, we have no way of knowing that the plaid path associated to our infinite orbit is fat. In the next section we give an alternate proof that does produce an infinite orbit with a fat associated path.

21.6 SKETCH OF AN ALTERNATE PROOF

Here we sketch how to extract Theorem 0.7 from the work in [**S1**]. We won't give a complete proof however.

Say that a *near Cantor set* is a set of the form $C - C'$ where C is a Cantor set and C' is a countable set. Let $I = (0, 2) \times \{1\}$. Let I_A denote the set of unbounded outer billiards orbits which intersect I for the irrational parameter A. In [**S1**] we proved that I_A is a near Cantor set. Moreover, we proved that I_A is dynamically minimal in the sense that the orbit of each $\zeta \in I_A$ intersects I_A in a dense set.

Referring to the graph PET $F_Y : Y \to Y$, the set $J_A \subset Y_A$ corresponding to I_A intersects the line segment J joining $(0, 0, 0, A)$ to $(1, 1, 1, A)$ in a near

Cantor set J_A. Looking at the formula for the linear part of the map Ω, given in Equation 17.8, we see that $\Omega^{-1}(J_A)$ contains a near Cantor set K_A that lies in a line segment K that is parallel to $(0, -1, 1, 0)$ and contained in a slice $\{x\} \times \boldsymbol{R}^2 \times \{P\}$, where $x \in \boldsymbol{Q}[P]$. From this we see that only countably many points of K_A fail to satisfy our definedness criterion above. Hence K_A has some well-defined orbits. By the Orbit Equivalence Theorem, these orbits are infinite.

Let $\zeta \in K_A$ and let $\Omega(\zeta) \in J_A$. The arithmetic graph γ associated to the orbit of $\Omega(\zeta)$ is fat because γ returns infinitely often to the 1-neighborhood of the line L of slope $-A$ through the origin, and also γ rises unboundedly far away from L.

We choose a sequence $\{A_n\}$ of even rationals converging to A. We let ζ_n be a sequence of points converging to ζ, with $\zeta_n \subset X_{A_n}$. The corresponding plaid polygons $\{\pi_n\}$ have π as a geometric limit. The corresponding arithmetic graph polygons $\{\gamma_n\}$ have γ as a geometric limit. Here γ_n is the arithmetic graph associated to $\Omega(\zeta_n)$. By construction, $\{\gamma_n\}$ exits every strip in the plane. Hence, by the Quasi-Isomorphism Theorem, so does $\{\pi_n\}$. Hence π is fat.

We showed in [**S1**] (among other things) that there is an infinite set $\{V_i\}$ of vectors such that γ contains a set of the form

$$\bigcup_\beta \left(\sum_j \beta_j V_j \right). \tag{21.4}$$

Just as in Equation 6, the union takes place over all finite binary sequences β, the sequence $\{V_j\}$ is unbounded, and the ordering on the points in the union coincides with the lexicographic ordering on the finite binary strings. Furthermore, all the points in Equation 21.4 lie within 1 unit of the Y-axis.

Given the Quasi-Isomorphism Theorem, the corresponding set vertices of π is quite close to a large-scale Cantor set. However, it might take some doing to show that it equals a large-scale Cantor set right on the nose. We will not pursue this, because we will give an independent proof.

Chapter Twenty-Two

Existence of Many Large Orbits

22.1 CHAPTER OVERVIEW

We call a plaid polygon *N-fat* if it is not contained in any strip of width N. As a related notion, we call a plaid polygon *N-long* if it has diameter at least N. Here is the result. In this chapter we will prove Theorem 0.8.

Theorem 22.1 *Let $\{p_k/q_k\} \subset (0,1)$ be any sequence of even rational numbers with a limit that is neither rational nor quadratic irrational. Let $\{B_k\}$ be any sequence of associated blocks. Let N be any fixed integer. Then the number of N-fat plaid polygons in B_k is greater than N provided that k is sufficiently large.*

Our proof relies on Theorem 0.7, which we have so far just proved using work in [**S1**]. As we mentioned in the last chapter we will give an independent proof of Theorem 0.7 later. For the reader who does not have the patience to read our independent proof, and who does not want to use the results of [**S1**], we observe the following. If we just rely on Theorem 21.1, then our proof below establishes the same result as Theorem 0.8 except with *N-long* in place of *N-fat*.

In §22.2 we study equidistribution properties of the plaid PET map Φ_A, as a function of A. In §22.3 we will use these equidistribution properties to show that the N-fat polygons essentially appear everywhere in the planar plaid model. We call the result the Ubiquity Lemma. The only thing we have to worry about is that all the N-fat polygons produced by the Ubiquity Lemma are really just a few giant polygons.

In §22.4 we examine how the plaid model interacts with the grid of all lines of capacity at most K. We show that sometimes these grid lines make impenetrable barriers that separate plaid polygons from each other. The main result is the Rectangle Lemma.

In §22.5 we use the Rectangle Lemma on many scales in order to show the existence of many distinct N-fat polygons. The Grid Supply Lemma below guarantees the existence of these barriers on many scales.

The remainder of the chapter is devoted to proving the Grid Supply Lemma. In §22.6 we discuss some properties of continued fractions and circle rotations. Finally, in §22.7 we prove the Grid Supply Lemma.

22.4 THE RECTANGLE LEMMA

In this section we show how the grid lines in the plaid model divide up the plaid polygons to a certain extent. This result will be a tool for the proof of Theorem 0.8. Fixing a parameter p/q, a block B, and an even integer $K \geq 0$ let Γ_K denote the union of all the lines of capacity at most K which intersect B. The set Γ_K divides B into a grid of $(K+1)^2$ rectangles.

Lemma 22.6 *There are at most $(K+1)^2/2 - 1$ light points on the intersection $\Gamma_K \cap B$.*

Proof: We know from Theorem 2.3 that a line of capacity k contains at most k light points. Since there are 4 lines of capacity k for each $k = 0, 2, ..., K$, this gives a total of

$$8 \sum_{k=1}^{K/2} k = (K+1)^2 - 1.$$

This completes the proof. $\square$

Call a rectangle *impermeable* if it intersects no plaid polygons.

Lemma 22.7 (Rectangle) *For all parameters and all blocks B and all choices of K, at least one of the rectangles of $B - \Gamma_K$ is impermeable.*

Proof: Suppose there are no impermeable rectangles and derive a contradiction. There are a total of $(K+1)^2$ rectangles. For the same reason as in the proof of Theorem 3.1, each rectangle in the partition has an even number of light points on its boundary. At the same time, each light point belongs to at most 2 rectangles. Hence there are at least $(K+1)^2$ light points. Hence, there are at least $(K+1)^2$ light points on $\Gamma_K \cap B$. This contradicts Lemma 22.6. $\square$

22.5 PROOF OF THE MAIN RESULT

Let N be as in Theorem 0.8. Say that a *fat disk* is a disk which satisfies the conclusions of the Ubiquity Lemma. This notion depends on N, but we fix N once and for all.

The Ubiquity Lemma says that every fat disk intersects an N-fat plaid polygon. When k is large there are many fat disks contained in B_k, but perhaps they all intersect the same N-fat plaid polygon. We want to use the Rectangle Lemma to separate many of these fat disks so that the N-fat polygons they intersect are distinct. We do this by applying the Rectangle Lemma many times, for different choices of the value K.

Assume for the moment that the parameter p/q is fixed. We call the grid Γ_K from the Rectangle Lemma *fat* if it has side length at least $2N$. Suppose that $K < L$, we say that the two grids Γ_K and Γ_L are *totally different* if no rectangle of Γ_K is a rectangle of Γ_L. What we mean is that every rectangle of Γ_K is nontrivially subdivided into rectangles of Γ_L. If we apply the Rectangle Lemma to Γ_K and Γ_L in this situation, we produce distinct empty rectangles which are either nested or have disjoint interiors. Following this section we will use some elementary number theory to establish the following result:

Lemma 22.8 (Grid Supply) *Let Ω be any positive integer. Once k is sufficiently large there are Ω pairwise totally different fat grids relative to p_k/q_k.*

We choose $\Omega = 4^N$. It follows from the Ramsey Theorem that given Ω empty rectangles in the plane, which are either nested or disjoint, there are either N pairwise disjoint rectangles in the collection or N mutually nested rectangles in the collection.

In the disjoint case, we can find N pairwise disjoint fat disks $D_1, ..., D_N$ which are separated from each other by empty rectangles. We can also do this in the nested case, though it is less obvious. The idea in the nested case is that the region $R_1 - R_2$, where R_1 and R_2 are two nested empty rectangles, is a union of other rectangles all having minimum side length $2N$. Hence $R_1 - R_2$ contains a fat disk.

By the Ubiquity Lemma, we can find an N-fat plaid polygon P_j which intersects D_j. The polygons $P_1, ..., P_N$ are pairwise disjoint because they are all separated from each other by empty rectangles. This completes the proof of Theorem 0.8 modulo the proof of the Grid Supply Lemma.

22.6 THE CONTINUED FRACTION LENGTH

Definition: Given any number $x \in [0,1]$, let $\Lambda(x)$ denote the length of the continued fraction of x. Note that $\Lambda(x)$ is finite if and only if x is rational.

It is fairly easy to see that $\Lambda(p_k/q_k) \to \infty$ when p_k/q_k has an irrational limit. The rational number which determines the geometry of the grids at the parameter p/q is $\widehat{\tau}/\omega$. Here $\omega = p + q$ and $\widehat{\tau} \in (0, \omega)$ satisfies $2p\widehat{\tau} \equiv 1$ mod ω. The sequence $\{\widehat{\tau}_k/\omega_k\}$ might converge to a rational number, or even 0. Nonetheless, we will show that $\Lambda(\widehat{\tau}_k/\omega_k) \to \infty$. Note that the sequence $\{2p_k/\omega_k\}$ also has an irrational limit. Using this observation, the result we seek is an immediate consequence of the following result.

Lemma 22.9 *Suppose $\{A_n/C_n\}$ is an infinite sequence of rational numbers in $(0,1)$ having an irrational limit. Suppose $A_n B_n \equiv 1$ mod C_n. Then $\Lambda(B_n/C_n) \to \infty$.*

Proof: Consider the two matrices

$$\alpha = \begin{bmatrix} 1 & 1 \\ 0 & 1 \end{bmatrix}, \qquad \beta = \begin{bmatrix} 1 & 0 \\ 1 & 1 \end{bmatrix}. \qquad (22.2)$$

These matrices generate $PSL_2(\mathbf{Z})$. Here is a well-known estimate on $\Lambda(p/q)$. We find any other rational p'/q' such that $\det(\mu) = 1$, where

$$\mu = \begin{bmatrix} p & p' \\ q & q' \end{bmatrix}. \qquad (22.3)$$

We write

$$\mu = \alpha^{n_1} \beta^{n_2} \alpha^{n_3} \beta^{n_4} ... \alpha^{n_k} \qquad (22.4)$$

in the shortest possible way. Then $|\Lambda(p/q) - k|$ is uniformly bounded.

We also note that $\Lambda(p/q)$ and $\Lambda(q/p)$ differ by at most 1. By definition, there is some integer K_n so that $A_n B_n - K_n C_n = 1$. But then $\Lambda(A_n/C_n)$ is comparable to $\Lambda(C_n/A_n)$, which is comparable to the word length of

$$M = \begin{bmatrix} C_n & B_n \\ A_n & K_n \end{bmatrix}. \qquad (22.5)$$

Note that α and β are transposes of each other: $\alpha = \beta^t$. But then the word length of M^t is the same as the word length of M. Finally, the word length of M^t is comparable to $\Lambda(B_n/C_n)$. $\square$

Slow Euclidean Algorithm: Given two positive integers a and b with $a < b$ we define the *slow Euclidean algorithm* for a, b as follows. We start with the pair $(A_0, B_0) = (a, b)$ and inductively define

$$A_{k+1} = \min(A_k, B_k - A_k), \qquad B_{k+1} = \max(A_k, B_k - A_k). \qquad (22.6)$$

We stop when we reach the pair $(A_n, B_n) = (d, d)$, where d is the greatest common divisor of a and b. We call two sets S, S' of positive integers *totally distinct* if $\min S > \max S'$. We mean to apply this definition to our pairs.

A certain subsequence of these pairs gives the successive terms of the ordinary Euclidean algorithm for the pair (a, b). If $(A, B), (A', B'), (A'', C'')$ are three consecutive terms corresponding to the ordinary Euclidean algorithm, then (A, B) and (A'', B'') are totally distinct. Hence, the number of totally distinct pairs in the list is at least $\Lambda(a/b)/2 - 2$. We have subtracted off 2 so that we don't have to think too hard about what happens for very small examples.

We can perform the same process on any pair of positive numbers $\rho < \sigma$, as long as ρ/σ is rational. In particular, if we take $\rho = p/q$ and $\sigma = 1$, then the number of totally distinct terms is at least $\Lambda(\rho)/2 - 2$. Compared to the previous process, all we are doing here is dividing every term in sight by q. Likewise if we take $\rho = p/q$ and $\sigma = 1 - \rho$ then the number of totally distinct terms is at least $\Lambda(\rho')/2 - 2$ where $\rho' = \rho/(1 - \rho)$. The two quantities $\Lambda(\rho)$ and $\Lambda(\rho')$ differ by at most 2. Hence, when we apply the process to the pair $(\rho, 1 - \rho)$ we produce at least $\Lambda(\rho)/2 - 4$ totally distinct pairs.

Circle Rotations: Now we describe a process very closely related to the slow Euclidean algorithm. Consider some $\rho = p/q \in (0,1)$. We distribute K points $x_0, ..., x_{K-1}$ in the circle $\boldsymbol{R}/\boldsymbol{Z}$ according to the formula

$$x_k = k\rho \bmod \boldsymbol{Z}. \tag{22.7}$$

Call this partition $I(K)$. Let $J(K)$ denote the set of sizes of intervals in $I(K)$. For instance $J(1) = \{\rho, 1-\rho\}$. We consider what happens for $K = 0, ..., q-1$. Here are some well-known properties of these partitions:

- $J(K)$ consists of at most 3 sizes.

- When $J(K)$ consists of 3 sizes, we have $J(K) = \{A, B, C\}$ with $A+B = C$. In this case, there is some larger K' such that $J(K') = \{A, B\}$.

- If $J(K) = \{A, B\}$ with $A < B$, then either $A = 1/q$ and $B = 2/q$ or there is some larger K' such that $J(K') = \{B - A, A, B\}$.

From these properties, we deduce the fact that every term in the slow Euclidean algorithm for $(\rho, 1 - \rho)$ appears as some $J(K)$.

From what we have said above, the number of totally distinct partitions is at least $\Lambda(\rho)/2 - 4$. If we restrict our attention to totally distinct partitions having an odd index – i.e., $J(2K - 1)$ – then we still get at least $\Lambda(\rho)/6 - 6$ totally distinct ones. We are imagining taking a list $K_0, ..., K_\ell$ of indices corresponding to totally distinct partitions, pruning the list to be $K_0, K_3, K_6, ...$, then replacing K_j by K_{j+1} if necessary to make it odd.

Modified Circle Rotations: Now we consider a modified process which is closer to the generation of the grids associated to the Grid Supply Lemma. We can distribute $2K$ points $x_0, y_0, ..., x_{K-1}y_{K-1}$ by setting x_j as in the circle rotation case and $y_j = 1 - x_j$. The resulting partition $I(K)$ has $2K$ intervals. Notice that the set of gaps in $I(K)$ is the same as the set $J(2K-1)$ considered above. Hence, the modified process produces at least $\Lambda(\rho)/6 - 6$ distinct partitions.

22.7 THE END OF THE PROOF

Now we return to the notation from the Grid Supply Lemma. Combining the work in the previous section, we see that

$$\lim_{k \to \infty} \Lambda(\widehat{\tau}_k / \omega_k) = \infty. \tag{22.8}$$

We construct the grid Γ_K by distributing the coordinate-axes intercepts of the lines according to the modified circle rotation process. This shows that, once k is large enough, there are at least $\Omega + N$ totally distinct grids. Since these grids are nested, the largest Ω grids will be N-fat. This completes the proof of the Grid Supply Lemma.

Chapter Twenty-Three

Infinite Orbits Revisited

23.1 CHAPTER OVERVIEW

The rest of the monograph is devoted to giving a self-contained proof of Theorem 0.7.

In §23.2 we describe a sequence of even rationals $\{p_n/q_n\}$ that converges to A. This is not the first sequence one would think of, but we found it by experimentation and it turns out to be very well adapted to the plaid model. Let π_n be the polygon from Theorem 3.1 that corresponds to p_n/q_n.

In §23.3 we state the two main technical results, the Box Theorem and the Copy Theorem. The Box Theorem says that a certain subset $\alpha_n \subset \pi_n$ is an arc. The Copy Theorem says that α_{n+1} contains two translated copies of α_n. When we iterate this result, we show that the set of vertices of α_n contained in the line $\{1/2\} \times \mathbf{R}$ contains a set $\langle \pi_n \rangle$ of points that are arranged like a large-scale Cantor set.

In §23.4 we show how to choose a sequence $\{c_n\}$, where $c_n \in \langle \pi_n \rangle$ is chosen so that the corresponding sequence $\{\Phi_{A_n}(c_n)\}$ converges to a point in X which satisfies our definedness criterion. The Cantor-set like structure of $\langle \pi_n \rangle$ is what makes this step possible. Finally, we take a geometric limit and complete the proof of Theorem 0.7.

In §23.5 we state three auxiliary results about arc copying in the plaid model. In §23.6 we deduce the Box Theorem from one of these auxiliary lemmas. In §23.7 we deduce the Copy Theorem from the auxiliary lemmas and some elementary number theory. Thus, after this chapter finishes, the only remaining task is to prove the auxiliary copy lemmas and prove a few lemmas in elementary number theory.

23.2 THE APPROXIMATING SEQUENCE

Here we construct a sequence of even rationals converging to A. If a/b is an even rational, then there is a unique even rational a'/b' with

$$|a'b - b'a| = 1, \qquad \omega' < \omega. \tag{23.1}$$

Here $\omega = a + b$ and $\omega' = a' + b'$. We call a'/b' the *even predecessor* of a/b.

Lemma 23.1 $A = \lim p_n/q_n$ *where* $\{p_n/q_n\}$ *is a sequence in which each term is the even predecessor of the next term.*

Proof: Let $\boldsymbol{H}^2$ denote the upper half-plane model of the hyperbolic plane. We have the Farey triangulation of $\boldsymbol{H}^2$ by ideal triangles. The geodesics bounding these triangles join rationals p_1/q_1 and p_2/q_2 such that

$$|p_1 q_2 - p_2 q_1| = 1.$$

We call these geodesics the *Farey geodesics*. We call a Farey geodesic *even* if it joins two even rationals.

The union E of even geodesics is invariant under the group $\Gamma_2 \subset PSL_2(\boldsymbol{Z})$ of matrices congruent to the identity mod 2. The quotient $\boldsymbol{H}^2/\Gamma$ is a 3-punctured sphere, and E/Γ_2 is a single even geodesic joining two of the cusps. From this we see that E is a tree and that the components of $\boldsymbol{H}^2 - E$ are neighborhoods of horodisks centered at odd rationals.

Let $\overrightarrow{E}$ denote the subset of E joining even rationals in $[0,1)$. We direct edges in $\overrightarrow{E}$ towards the rationals with larger denominators. This makes $\overrightarrow{E}$ a directed, embedded, planar tree with initial node $0/1$. From this structure, there is a unique infinite path in $\overrightarrow{E}$ which joins $0/1$ to A, and it is the limit of finite paths in $\overrightarrow{E}$ which join $0/1$ to any sequence of even rationals that limits to A. $\square$

Lemma 23.2 *Let $\{p_n/q_n\}$ be the even predecessor sequence which converges to A. There are infinitely many values of n such that $2\omega_n < \omega_{n+1}$.*

Proof: If this lemma is false, then we might chop off the beginning and assume that this never happens. We never have $2\omega_n = \omega_{n+1}$ because ω_{n+1} is odd. So, we always have $2\omega_n > \omega_{n+1}$. In this case, define

$$p'_{n-1} = 2p_n - p_{n+1}, \qquad q'_{n-1} = 2q_n - q_{n+1}.$$

Note that p'_{n-1}/q'_{n-1} must be the even successor of p_n/q_n. Therefore, we have $p'_{n-1} = p_{n-1}$ and $q'_{n-1} = q_{n-1}$. This means that

$$p_n = \frac{p_{n-1} + p_{n+1}}{2}, \qquad q_n = \frac{q_{n-1} + q_{n+1}}{2}.$$

The sequences $\{p_n\}$ and $\{q_n\}$ are therefore linear progressions and the co-efficients describing these progessions are rational. But then A is rational. This is a contradiction. $\square$

Suppose that p_1/q_1 is the even predecessor of p_2/q_2. Even though p_1/q_1 is the unique even predecessor of p_2/q_2, there are other even rationals p'_2/q'_2 having p_1/q_1 as an even predecessor. We define the *core predecessor* of p_2/q_2 to be the even rational with the smallest denominator that has p_1/q_1 as an even predecessor. Thus $2/5$ is the core of every even rational having $1/2$ as its even predecessor.

We define the *predecessor sequence* to be the sequence obtained from the even predecessor sequence by inserting the core of every term and then re-moving repeaters. The example below shows part of the core sequence. The terms in brackets do not belong to the even predecessor sequence.

$$\frac{1}{2}, \; \boxed{\frac{\mathbf{2}}{\mathbf{5}}}, \; \frac{4}{9}, \; \frac{7}{16}, \; \boxed{\frac{\mathbf{10}}{\mathbf{23}}}, \; \frac{24}{55}, \; \frac{\mathbf{55}}{\mathbf{126}}, \ldots$$

Finally, we define a term p_n/q_n in the predecessor sequence as *special* if either

- p_n/q_n is not in the even approximating sequence, or

- p_n/q_n is in the even predecessor sequence and $2\omega_n < \omega_{n+1}$. Here $\omega_n = p_n + q_n$.

The bold terms in our example are special.

If follows from Lemma 23.2 that there are infinitely many special terms in the core sequence. We call the sequence of special terms the *approximating sequence*.

23.3 THE COPY THEOREM

All the definitions in this section are defined relative to the approximating sequence $\{p_n/q_n\}$. For any object R, we let R_n denote the object defined relative to the parameter p_n/q_n.

Let B_n denote the 0th block and let $R_n \subset B_n$ denote the rectangle bounded by the top, bottom, and left sides of B_n, and the vertical segment in B_n of capacity at most 4 that lies as close as possible to the left side of B_n. Let π_n be the big polygon from Theorem 3.1. Let $\alpha_n = \pi_n \cap R_n$. In subsequent chapters we will prove the following result.

Lemma 23.3 (Box) α_n *is a single arc.*

We define $\eta_n = \omega_n - 2\tau_n$. Here ω_n and τ_n are as in §1.2. The quantity η_n measures the distance between the horizontal lines of capacity 2 in B_n. Here is our main technical result.

Theorem 23.4 (Copy) *Let p_n/q_n and p_{n+1}/q_{n+1} be successive terms in the approximating sequence. There are vertical translations $\Upsilon_{B,n}$ and $\Upsilon_{T,n}$ such that $\Upsilon_{B,n}(\alpha_n)$ and $\Upsilon_{T,n}(\alpha_n)$ are both arcs of α_{n+1}, symmetrically placed with respect to the horizontal midline of B_{n+1}. The image $\Upsilon_{B,n}(\alpha_n)$ lies below the horizontal midline of B_{n+1} and $\Upsilon_{T_n}(\alpha_n)$ lies above it. Finally,*

$$\Upsilon_{T,n}(*) - \Upsilon_{B,n}(*) = (0, \eta_{n+1} \pm \eta_n).$$

Figure 23.1 shows the Copy Theorem in action for the two parameters $4/9$ and $9/20$. Here we are just showing $\Upsilon_B(\alpha_{4/9})$ inside $\alpha_{9/20}$. The arc $\Upsilon_T(\alpha_{4/9})$ is symmetrically placed at the top of the picture. The dark horizontal lines in the figure are the lines of capacity 2. Note how Υ_B lines them up.

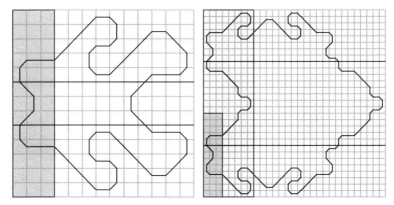

Figure 23.1: Arc copying $p/q = 4/9$ and $p/q = 9/20$.

The left-hand side of Figure 23.2 shows what the Box Lemma and the Copy Theorem imply about the structure of α_n. The curve inside the large rectangle is α_n and the curves inside the smaller rectangles are copies of α_{n-1}. The right-hand side of Figure 23.2 shows what the Box Lemma and 2 applications of the Copy Theorem imply about α_n. One can see how this idea might be iterated.

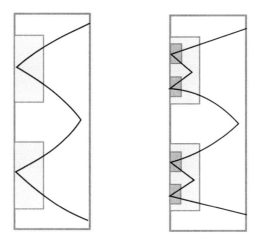

Figure 23.2: Iterating the Copy Theorem.

Let $\langle \pi_n \rangle$ denote the set of vertices of α_n having first coordinate $1/2$. We call $\langle \pi_n \rangle$ the *anchor* of π_n.

Corollary 23.5 $\langle \pi_n \rangle$ *contains* 2^k *points of the form*

$$(1/2, 1/2) + (0, y_n) + \sum_{i=0}^{k-1} \epsilon_i(0, d_i),$$

where d_i *is one of the two terms* $\eta_{i+1} \pm \eta_i$ *for each* i. *Here* y_n *is some integer and* $\{\epsilon_i\}$ *is any binary sequence of length* k.

23.4 THE END OF THE PROOF

Any infinite binary sequence β gives rise to a sequence $\{c_n\}$ of points, where c_n is a vertex of $\langle \pi_n \rangle$. We simply take the sum in Corollary 23.5, taking $\epsilon_1, ..., \epsilon_n$ to be the first n bits of β. We call β *generic* if β has infinitely many 0s and infinitely many 1s. There are uncountably many generic sequences.

Lemma 23.6 *There exists a generic sequence β so that, for the corresponding sequence, $\{c_n\}$, the sequence $\{\Phi_n(c_n)\}$ has an accumulation point in X with a well-defined orbit.*

Proof: For convenience, we compute modulo Λ'_X. If our definedness criterion is satisfied when we reduce mod Λ'_X, it is also satisfied when we reduce mod Λ_X. Also for convenience, we drop the 4th coordinate from our calculation. First of all,

$$\Phi_n(0, y_n) = (2y_n, 0, 2P_n y_n) \equiv (0, -\mu_n, \mu_n) \mod \Lambda'_X. \tag{23.2}$$

We will compute below that

$$\Phi_n(0, d_n) = (0, -\theta_n, \theta_n), \qquad |\theta_n| < \frac{3}{2^n}. \tag{23.3}$$

Combining these equations with Corollary 23.5, we see that the sets $\Phi_n(\langle \pi_n \rangle)$ have a geometric limit which contains a Cantor set. To make this more clear, we can pass to a subsequence on which μ_n converges.

The geometric limit of $\Phi_n(\langle \pi_n \rangle)$ is uncountable and cannot lie entirely in the set $Q[P]$. There are uncountably many generic choices of the binary sequence β so that $\lim \Phi_n(c_n)$ is a point of this Cantor set not in $Q[P]$. Such a point has a well-defined orbit, by our definedness criterion.

Now we give the calculation. We compute

$$\Phi_n(0, \eta_n) = (0, -P_n \eta_n, P_n \eta_n). \tag{23.4}$$

We have

$$\eta_n P_n = (\omega_n - 2\tau_n) P_n = 2p_n - \frac{4\tau_n p_n}{\omega_n} = 2p_n - 2a_n - \frac{2}{\omega_n}.$$

Here a_n is some integer. This means that $P_n \eta_n \equiv 2/\omega_n \mod 2$. In other words

$$\Phi_n(\eta_n) = (0, -\theta'_n, \theta'_n), \qquad |\theta'_n| \leq \frac{2}{\omega_n} < \frac{2}{2^n}. \tag{23.5}$$

The last estimate comes from the fact, as seen by the conclusion of the Copy Theorem, that $\omega_{i+1} > 2\omega_i$ for all i. Finally, Equation 23.3 follows from the triangle inequality and from the fact that $d_n = \eta_{n+1} \pm \eta_n$, so that $|\theta_n| \leq (1 + 1/2)|\theta'_n|$. $\square$

Let π'_n denote the translated copy of π_n so that c_n moves to $(1/2, 1/2)$. We pass to a subsequence on which $\Phi_n(c_n)$ converges to a point with a well-defined orbit O. By Lemma 21.4, the plaid path π_∞ associated to O is the geometric limit of $\{\pi'_n\}$. There are infinitely many indices k for which some initial portion of π_∞ agrees up to translation with $\pi_k \cap R_k$. Here R_k is the box from the Copy Theorem. But this means that the projection of π_∞ onto the x-axis is unbounded. At the same time, $\langle \pi_\infty \rangle$ is infinite, being a geometric limit of the sets $\langle \pi'_n \rangle$. These two properties together imply that π_∞ is fat. This completes the proof of Theorem 0.7.

23.5 THE COPY LEMMA

Now we state three auxiliary lemmas about arc copying. We first classify pairs $(p'/q', p/q)$ of consecutive rationals which can appear in the core predecessor sequence.

- The pair is *strong* if p/q is its own core and p'/q' is the even predecessor of p/q and $2\omega' < \omega$.

- The pair is *weak* if p/q is its own core and p'/q' is the even predecessor of p/q and $2\omega' > \omega$.

- The pair is *core* if p'/q' is the core of p/q.

When p'/q' is in the approximating sequence, the pair is either strong or core.

In each of the three cases, we define a transformation Υ associated to the pair. In the strong and weak cases, Υ is the identity. In the core case, Υ is the vertical translation which maps the horizontal midline of R' to the horizontal midline of R. That is, Υ is vertical translation by $(\omega - \omega')/2$.

To each rational p/q we associate two rectangles. The rectangle $R = R_{p/q}$ is as above. The rectangle R^* is the largest sub-rectangle of R lying below the diagonal line $x + y = \omega$. We have $R^* \subset R$. We also associate two rectangles (S', R^*) to our pair $(p'/q', p/q)$. When the pair is weak we set $S' = (R')^*$. Otherwise we set $S' = R'$.

Let PL and PL' denote the union of all the plaid polygons for each of the parameters p'/q' and p/q. Let H'_2 and H_2 denote the union of the two capacity 2 horizontal lines relative to p/q and p'/q' respectively. Let LH_2 denote the lower of these two lines. Here is our main technical result.

Lemma 23.7 (Copy) *For any pair $(p'/q', p/q)$ of successive rationals in any core predecessor sequence, we have*

$$\Upsilon(S') \subset R^*, \qquad \Upsilon(H'_2) \cap LH_2 \neq \emptyset, \qquad \Upsilon(PL' \cap S') = PL \cap \Upsilon(S').$$

We only care about the Copy Lemma as it relates to the big polygons π' and π. This is what our pictures show. Figure 23.1 (also) illustrates the Copy Lemma for the pair $(4/9, 9/20)$ except that the right side shows $R_{9/20}$ rather than the smaller $R^*_{9/20}$. Figures 23.3-23.5 trace our 3 successive applications of the Copy Lemma to our example sequence above.

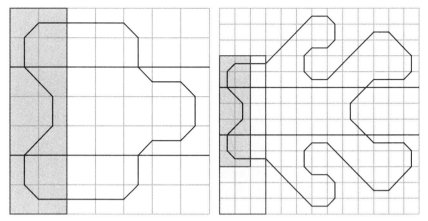

Figure 23.3: Core copying: $p'/q' = 2/5$ and $p/q = 4/9$.

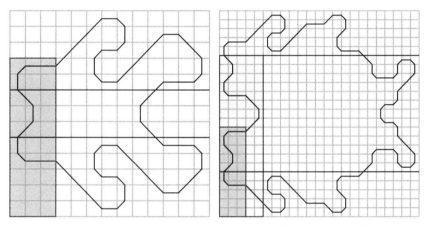

Figure 23.4: Weak copying: $p'/q' = 4/9$ and $p/q = 7/16$.

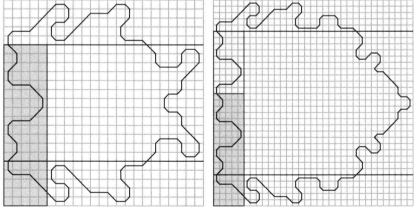

Figure 23.5: Weak copying: $p'/q' = 7/16$ and $p/q = 10/23$.

If we string the three applications of the Copy Lemma together, we see that

$$\Upsilon_B(\alpha_{2/5}) \subset \alpha_{10/23}, \tag{23.6}$$

where Υ_B is one of the two translations in the Copy Theorem. The map Υ_T is symmetrically defined. This gives the statement of the Copy Theorem for the pair $(2/5, 10/23)$.

Remark: Notice that our analysis misses some of the structure. We only pick up the fact that $\alpha_{10/23}$ contains 2 copies of $\alpha_{2/5}$ and not 3 copies. We do not pick up the middle copy. In the case above, one can deduce the existence of this middle copy from symmetry. In general, we would need to formulate the Copy Lemma more carefully to account for the complete structure of the union of vertices mentioned in Theorem 0.7.

Computer Tie-In: The subdirectory **CopyLemmas** contains a small version of the main program which is dedicated to illustrating the Copy Lemma. When you run this program, you see a more elaborate parameter entry system and two windows showing the plaid model. When you press one of the three buttons labeled *weak*, *strong*, or *core*, the computer generates a random example of a pair of rationals having this type. The two picture windows then show versions of the figures above for those parameters.

23.6 PROOF OF THE BOX THEOREM

Let p/q be an even rational parameter and let π be the polygon from Theorem 3.1. Let R be the rectangle associated to p/q. Three sides of R lie in the boundary of the block $[0, \omega]^2$. The polygon π does not intersect these three sides. The remaining side of R is the vertical line of capacity at most 4 that is closest to the left side of the block. We want to show that $\alpha = \pi \cap R$ is a single arc. It is enough to show that π intersects the right side ρ of R exactly twice.

By the analysis in Theorem 3.1, we know that π intersects ρ at least twice. We also know that π intersects ρ an even number of times, because π is a closed loop. Finally, we know that π intersects ρ at most 4 times, because ρ has capacity at most 4. If ρ has capacity 2, then π can intersect ρ at most twice. So, in this case, π must intersect ρ exactly twice. It remains to consider the case when ρ has capacity 4.

By induction, we can assume that p/q is the rational with the smallest sum $p + q$ for which we do not yet know the result. The left side of Figure 23.6 shows what would potentially go wrong. The grey lines have capacity 2 and the black line has capacity 4. The shaded rectangle is R. Given the way that π intersects the two horizontal grey lines, the only way π could intersect ρ four times is the way shown in Figure 23.6.

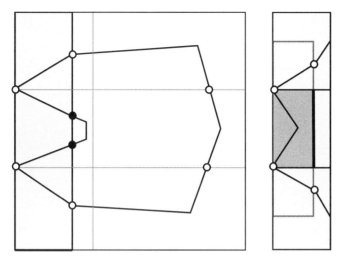

Figure 23.6: An extra intersection.

We will prove momentarily that p/q is not its own core. Thus, we can apply the Copy Lemma to the pair $(p'/q', p/q)$, where p'/q' is the core of p/q. Since we are in the core case, we have $S' = R'$. We now mention 3 facts.

1. By induction π' intersects the right boundary ρ' of R' exactly twice.

2. $\pi' \cap \rho'$ lies outside the band of $[0, \omega']^2$ bounded by the lines of H_2'.

3. By the Copy Lemma and symmetry, $\Upsilon(H_2') = H_2$.

The right-hand side of Figure 23.6 shows $\Upsilon'(R')$. The darkly shaded part is the portion between the lines of H_2. From Item 2, π' does not intersect ρ' between the two lines of H_2'. But then $\pi' \cap R' = \alpha'$, and α' is a single arc. Next, note that $\alpha \cap H_2 = \Upsilon'(\alpha' \cap H_2')$. Since $\Upsilon(\alpha') \subset \alpha$, we see that the portion of π between the points of $\pi \cap H_2$ must be the arc $\Upsilon(\alpha')$. By construction, this arc does not exit $\Upsilon(R')$, a subset of R. Hence the portion of α between the two points of $\alpha \cap H_2$ cannot exit R. This rules out the left side of Figure 23.6.

To finish the proof, we have to explain why p/q is not its own core. Let $\widehat{\tau}$ and τ be as in Lemma 1.2. Here $\tau = \min(\widehat{\tau}, \omega - \widehat{\tau})$. By Lemma 1.1 the lines of capacity 2 have the form $y = y_0$ with

$$y_0 \in \{\tau, \omega - \tau, 2\tau, \omega - 2\tau\}.$$

When ρ has capacity 4 it means that $\omega - 2\tau < \tau$, which is to say that $\omega < 3\tau$. Lemma 24.2, proved in the next chapter, says that p/q is not its own core.

23.7 PROOF OF THE COPY THEOREM

Here we will prove the Copy Theorem modulo a few statements about elementary number theory that we will establish in the next chapter. The proof is really just a generalization of the pictorial tour we gave using Figures 23.3-23.5. The hardest part is keeping the indices straight. To help with the indices, we denote rationals in the approximating sequence with bold letters. Two consecutive members $\mathbf{p_n}/\mathbf{q_n}$ and $\mathbf{p_{n+1}}/\mathbf{q_{n+1}}$ in the approximating sequence correspond to a fragment of the form

$$\mathbf{p_n}/\mathbf{q_n} = p_\ell/q_\ell, ..., p_m/q_m = \mathbf{p_{n+1}}/\mathbf{q_{n+1}} \tag{23.7}$$

in the core predecessor sequence. Note that there cannot be two terms in a row that do not belong to the even predecessor sequence, because they would then have the same even predecessor, and each would be the core of the other. Hence $m \geq \ell + 2$.

By definition, the pair $(p_\ell/q_\ell, p_{\ell+1}/q_{\ell+1})$ is not weak and the remaining pairs are weak. Since the first pair is not weak, the associated rectangle S_ℓ from the Copy Lemma is all of R_ℓ. Successive applications of the Copy Lemma now tell us that

$$\Upsilon_B(\alpha_\ell) \subset \alpha_m, \qquad \Upsilon_B(H_{2,\ell}) \cap LH_{2,m} \neq \emptyset. \tag{23.8}$$

Here $H_{2,k}$ is the set of horizontal lines of capacity 2 associated to the parameter p_k/q_k. Here Υ_B is the composition of all the translations appearing in the Copy Lemma as it is applied to the successive pairs in our sequence. The translation Υ_T is defined as

$$\Upsilon_T = \phi_m \circ \Upsilon_B \circ \phi_\ell. \tag{23.9}$$

Here ϕ_k is the reflection in the horizontal midline of the 0th block associated to p_k/q_k. By symmetry, Υ_T maps one of the lines of $H_{2,\ell}$ to the upper line of $H_{2,m}$. The line-matching properties of Υ_B and Υ_T are responsible for the formula for the translation distance $\Upsilon_T - \Upsilon_B$.

The only thing left to check is that $\Upsilon_B(S_\ell)$ lies below the horizontal midline of the block $[0, \omega_m]^2$. Note that all the nontrivial translation part of Υ_b comes from the first application of the Copy Lemma.

Case 1: Suppose that the first pair is strong. In this case, Υ_B is the identity. We have

$$\omega_\ell < \frac{1}{2}\omega_{\ell+1} < \omega_m.$$

Hence $\Upsilon_B(R_\ell)$ lies in the lower half of $[0, \omega_m]^2$.

Case 2: Suppose that the first pair is core. In this case, Υ_B maps the lower line of $H_{2,\ell}$ to the lower line of $H_{2,m}$. The distance from the lower line of $H_{2,k}$ to the top of R_k is $\omega_k - \tau_k$. Therefore, the top of $\Upsilon_B(S_\ell)$ is contained in the line

$$y = \tau_m + (\omega_\ell - \tau_\ell).$$

We just need to show that $2y < \omega_m$. This is the same as

$$2\omega_\ell - 2\tau_\ell < \omega_m - 2\tau_m.$$

Recall that $\eta_k = \omega_k - 2\tau_k$. So, our inequality becomes

$$\omega_\ell + \eta_\ell < \eta_m. \qquad (23.10)$$

It follows immediately from Statement 2 of Lemma 24.1, proved in the next chapter, that the sequence $\{\eta_k\}$ is non-decreasing. So, to prove our inequality it suffices to show that

$$\omega_\ell + \eta_\ell < \eta_{\ell+2}. \qquad (23.11)$$

Lemma 24.3, proved in the next chapter, establishes this inequality.

23.8 HIDDEN SYMMETRIES

Here we discuss the additional connection between the plaid model and the Truchet tiling. The result is an equality between portions of two different slices of the 3D plaid model. We will state the simplest version of what seems to be true. The interested reader can play with the program and see that more is true, at least for many parameters.

Let L denote the infinite strip $[0,1] \times \boldsymbol{R}$. We think of L as the *left edge*.

Conjecture 23.8 *The intersection of L with the plaid tiling at the parameter p/q agrees up to translation with the intersection of L with the vertical pixelated spacetime diagram of capacity $2p$.*

Figure 23.7 shows this for the parameter $4/9$.

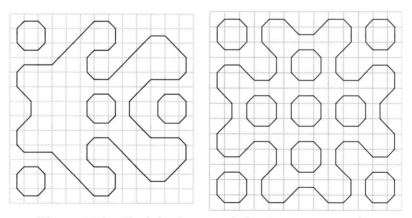

Figure 23.7: The left edges match for the parameter $4/9$.

Computer Tie-In: On the main program, open up the *planar* and *spacetime* popup windows. Also open up the *spacetime diag* control panel. On this control panel turn off everything but the *background*, *grid*, and *pixelated V* features. Use the arrow keys to set the capacity to $2p$, where p/q

is the parameter. Click the third or fourth square on the row of squares (underneath the letter V) to activate the picture of the capacity $2p$ vertical slice. In the *spacetime diag* window, use the keys a and s to shift the picture up and down until the left edge matches the left edge of the plaid tiling shown in the *planar* window. You can pick random parameters and try this experiment over and over again.

One plausible explanation for this is that the copying mechanism works the same way in both slices. Thus, if we can establish the equality we are talking about for the first few parameters, then the copying takes care of the rest. In order to really prove this, we would need to use stronger versions of the Copy Lemma so that it accounted for all the points on the left edge of the plaid tiling. Following this, we would have to prove a version of the Copy Lemma for the spacetime diagrams. This would surely work, but our proof of the Copy Lemma above is hard enough to discourage us.

Another plausible explanation is that this phenomenon comes from the hidden symmetries discussed in §5.6. However, we don't understand this well enough to give a proof along these lines.

Chapter Twenty-Four

Some Elementary Number Theory

24.1 CHAPTER OVERVIEW

In this chapter we prove some number-theoretic results about the sequences we defined in the previous chapter. In §24.2 we prove Lemma 24.1, a multipart structural result. In §24.3 we take care of the several number-theoretic details left over from §23.6 and §23.7.

24.2 A STRUCTURAL RESULT

First we gather together some definitions. We assume $p > 1$.

- $\omega = p + q$.

- $\tau \in (0, \omega/2)$ is the unique solution to $2p\tau \equiv \pm 1 \bmod \omega$.

- $\kappa \in \{0, 1, 2, ...\}$ is such that

$$\frac{\kappa}{2\kappa + 1} \le \frac{\tau}{\omega} < \frac{\kappa + 1}{2(\kappa + 1) + 1}. \tag{24.1}$$

- p'/q' is the even predecessor of p/q. That is, $|pq' - qp'| = 1$ and $\omega' < \omega$.

- $\overline{p}/\overline{q}$ is the core predecessor of p/q.

We mean to define ω, τ, κ for the other parameters as well. Thus, for instance, $\omega' = p' + q'$.

Lemma 24.1 *The following is true.*

1. *$\overline{p} = p - 2\kappa p'$ and $\overline{q} = p - 2\kappa q'$.*

2. *$\omega' = \omega - 2\tau = \overline{\omega} - 2\overline{\tau}$.*

3. *Suppose $\kappa = 0$. If $2\omega' > \omega$ then $\tau' = \tau$. Otherwise $\tau' + \tau = \omega'$.*

4. *$\frac{1}{1+2\kappa}\omega \le \overline{\omega} \le \frac{3}{3+2\kappa}\omega$.*

5. *$\overline{\omega} = (1 - 2\kappa)\omega + 4\kappa\tau$. If $\kappa = 1$ then $\overline{\omega} < 2\tau$.*

6. *$2p\tau \equiv 1 \bmod \omega$ iff $2\overline{p}\,\overline{\tau} \equiv 1 \bmod \overline{\omega}$.*

7. $4p\bar{\tau} \equiv \pm(4\kappa + 2) \bmod 2\omega$. *The sign is* 1 *iff* $2\tau \equiv 1 \bmod \omega$.

Proof: We first give a formula for p'/q'. There is some integer $\theta > 0$ so that

$$2p\tau = \theta(p + q) \pm 1. \tag{24.2}$$

Since $q > 1$ we have $\theta < 2\tau$. Since $\tau < (p + q)/2$ and $p > 1$ we have $\theta < p$. Hence $\theta < \min(p, 2\tau)$. Rearranging Equation 24.2 we get

$$|pq'' - qp''| = 1, \qquad p'' = \theta, \qquad q'' = 2\tau - \theta.$$

This implies that p''/q'' is in lowest terms. Moreover, both p'' and q'' are odd. Hence

$$p' = p - p'' = p - \theta, \qquad q' = q - q'' = q - 2\tau + \theta. \tag{24.3}$$

This gives us

$$\omega' = \omega - 2\tau \le \omega - 2\omega \frac{\overline{\kappa}}{2\kappa + 1} = \frac{\omega}{2\kappa + 1}. \tag{24.4}$$

Rearranging this, we get

$$2\kappa p' + 2\kappa q' + (p' + q') \le p + q. \tag{24.5}$$

Given that $|pq' - qp'| = 1$, Equation 24.5 forces $2\kappa p' < p$ and $2\kappa q' < q$.

Statement 1: Now we define $p^* = p - 2\kappa p'$ and $q^* = q - 2\kappa q'$. We will establish that p^*/q^* is the core predecessor of p/q. By Equation 24.5, both p^* and q^* are positive. By construction, p^*/q^* and p'/q' are Farey related. Hence $p^*/q^* \in (0, 1)$ and p^*/q^* is written in lowest terms. Equation 24.4 says that $p^* \ge p'$ and $q^* \ge q'$. If $p^* = p'$ we also have $q^* = q'$ because p'/q' and p^*/q^* are Farey related. Likewise, if $q^* = q'$ then $p^* = p'$. If $p^* = p'$ and $q^* = q'$ then $p = (2\kappa + 1)p'$ and $q = (2\kappa + 1)q'$, contradicting the fact that p and q are relatively prime. Thus we have $p' < p^*$ and $q' < q^*$. Hence p'/q' is the even predecessor of p^*/q^*.

If p^*/q^* is not the core predecessor of p/q then we can repeat the construction above with p^*/q^* in place of p/q, defining $p^{**} = p^* - 2\kappa^* p' > 0$ and $q^{**} = q^* - 2\kappa^* q' > 0$. But this leads, via the rearrangement of Equation 24.4, to the conclusion that

$$\frac{\tau}{\omega} > \frac{\kappa + \kappa^* + 1}{2(\kappa + \kappa^* + 1) + 1}.$$

This contradicts the definition of κ.

Statement 2: Since Equation 24.4 applies both to the pair $(p'/q', p/q)$ and to the pair $(p'/q', \bar{p}/\bar{q})$, we see that $\omega' = \omega - 2\tau = \bar{\omega} - 2\bar{\tau}$. This is Statement 2.

Statement 3: Assume $\kappa = 0$. Let θ be as above. Choosing the sign as in Equation 24.2, we have

$$\theta\omega' \pm 1 = \theta(\omega - 2\tau) \pm 1 = (\theta\omega \pm 1) - 2\theta\tau = 2p\tau - 2\theta\tau = 2p'\tau. \tag{24.6}$$

This shows that $2p'\tau \equiv \pm 1 \mod \omega'$. But, by definition, $2p'\tau' \equiv \pm 1 \mod \omega'$. Hence $2p'\tau \equiv \pm 2p'\tau' \mod \omega'$. Since $2p'$ is relatively prime to ω' we see that

$$\tau \equiv \pm\tau' \mod \omega'. \tag{24.7}$$

We compute

$$\omega' - \tau = \omega - 3\tau > 0.$$

The last inequality comes from the fact that $\kappa = 0$. Equation 24.7 now tells us that either $\tau' = \tau$ or $\tau + \tau' = \omega'$. It remains to sort out these two cases.

Observe first that $\tau' < \omega'/2$. So, in either case we have $\tau' \leq \tau$. Suppose first that $2\omega' > \omega$. By Statement 2 we have

$$2\omega - 4\tau = 2\omega' > \omega.$$

This leads to $4\tau < \omega$. If $\tau + \tau' = \omega'$ then

$$\omega < 2\omega' = 2\tau' + 2\tau \leq 4\tau < \omega.$$

This is a contradiction. Hence $\tau = \tau'$ in this case.

Suppose now that $2\omega' < \omega$. This leads to $4\tau > \omega$. But if $\tau' = \tau$ in this case, we have

$$\omega > 4\tau = 4\tau' < 2\omega' < \omega,$$

and we have a contradiction. Hence $\tau' + \tau = \omega'$ in this case.

Statement 4: We have

$$\overline{\omega} = \omega - 2\kappa\omega' = \omega - 2\kappa(\omega - 2\tau) = (1 - 2\kappa)\omega + 4\kappa\tau. \tag{24.8}$$

We have

$$\frac{\kappa}{2\kappa + 1} \leq \frac{\tau}{\omega} < \frac{\kappa + 1}{2\kappa + 3}. \tag{24.9}$$

Dividing Equation 24.8 by ω, we get

$$\frac{\overline{\omega}}{\omega} = 1 - 2\kappa + 4\kappa\frac{\tau}{\omega} \leq 1 - 2\kappa + 4\kappa\frac{\kappa + 1}{2\kappa + 3} = \frac{3}{3 + 2\kappa}. \tag{24.10}$$

Similarly

$$\frac{\overline{\omega}}{\omega} = 1 - 2\kappa + 4\kappa\frac{\tau}{\omega} \geq 1 - 2\kappa + 4\kappa\frac{\kappa}{2\kappa + 1} = \frac{1}{1 + 2\kappa}. \tag{24.11}$$

When $\kappa = 1$ we get

$$\overline{\omega} = 4\tau - \omega = (2\tau - \omega) + 2\tau < (2\tau - \omega) + \omega = 2\tau.$$

The first equality is Equation 24.8. We also used the fact that $\tau < \omega/2$.

Statement 5: This is just Equation 24.8.

Statement 6: We will suppose that $2p\tau \equiv 1 \mod \omega$ and $2\overline{p}\overline{\tau} \equiv -1 \mod \overline{\omega}$ and derive a contradiction. From Statement 2 we get $p\omega' \equiv -1 \mod \omega$. Hence $p\omega' - a\omega = -1$ for some $a \in \mathbf{Z}$. But $p\omega' - p'\omega = \pm 1$. From this we conclude that either $a = p'$ or else $(a - p')\omega = 2$. The latter case is absurd. Hence $a = p'$. In short,

$$p\omega' - p'\omega = -1.$$

The same argument shows that

$$\overline{p}\omega' - p'\overline{\omega} = +1.$$

Subtracting these equations and using Statement 1, we get

$$2 = (p - \overline{p})\omega' - p'(\overline{\omega} - \omega) = (2\kappa p')\omega' - p'(2\kappa\omega') = 0.$$

This is a contradiction.

Statement 7: We assume that $2p\tau \equiv +1 \bmod \omega$. The (-1) case has the same treatment. Combining Statements 1 and 2 we get

$$2\overline{\tau} = \omega - (2k + 1)\omega'. \tag{24.12}$$

By Statement 2 we have $\omega' = \omega - 2\tau$. Hence

$$p\omega' = p\omega - 2p\tau \equiv -1 \bmod \omega.$$

Combining this with Equation 24.12, we get

$$2p\overline{\tau} \equiv +(2k + 1) \bmod \omega.$$

Multiplying everything in sight by 2 we get Statement 7. $\square$

24.3 UNFINISHED BUSINESS

In this section we take care of the number-theoretic details left over from our proof of the Box Theorem and the Copy Theorem. We keep the notation from the previous section. Again, we assume $p > 1$. Recall that $\eta = \omega - 2\tau$.

Lemma 24.2 $3\tau < \omega$ if and only if $\kappa = 0$ if and only if p/q is its own core.

Proof: The equation $\omega = 3\tau$ yields $\pm 1 = 3k\tau + 2p\tau$ for some k, which forces $\tau = 1$ and $\omega = 3$. So, either $\omega < 3\tau$ or $\omega > 3\tau$. If follows from Equation 24.1 that $\omega < 3\tau$ if and only if $\kappa > 0$. If follows from Statement 1 of Lemma 24.1 that $\kappa = 0$ if and only if p/q is its own core. $\square$

Lemma 24.3 Let $p_0/q_2, p_1/q_1, p_2/q_2$ be three distinct even rationals such that p_1/q_1 is the even predecessor of p_2/q_2 and p_0/q_0 is the core predecessor of p_1/q_1. Then $\omega_0 + \eta_0 < \eta_2$.

Proof: By Statement 2 of Lemma 24.1 we have $\omega_1 = \eta_2$. So, our inequality is the same as $\omega_0 + \eta_0 < \omega_1$. We change notation so that $p/q = p_1/q_1$ and

$\overline{p}/\overline{q} = p_0/q_0$. We want to show that $\overline{\omega} + \overline{\eta} < \omega$. We have $\kappa \geq 1$. If $\kappa > 1$ then by Statement 4 of Lemma 24.1 we have

$$\overline{\omega} + \overline{\eta} = 2\overline{\omega} - 2\overline{\tau} < 2\overline{\omega} < \omega.$$

If $\kappa = 1$ we have

$$\overline{\omega} + \overline{\eta} = 2\overline{\omega} - 2\overline{\tau} = \overline{\omega} + (\overline{\omega} - 2\overline{\tau}) = \overline{\omega} + (\omega - 2\tau) < 2\tau + (\omega - 2\tau) = \omega.$$

Here we have used Statement 2 and Statement 4 of Lemma 24.1. $\square$

Chapter Twenty-Five

The Weak and Strong Case

25.1 CHAPTER OVERVIEW

In this chapter we complete the proof of the weak and strong case of the Copy Lemma. The two cases have just about the same proof. We fix the parameter $A = p/q$ and define $P = 2A/(1 + A)$ as usual. We make the same definitions for the even predecessor $A' = p'/q'$. Our notation convention is that for any object X we define in terms of p/q, the object X' is defined in terms of p'/q'.

In §25.2 we prove the first two statements of the Copy Lemma. The rest of the chapter is devoted to proving the third statement.

In §25.3 we prove an easy technical lemma.

In §25.4 we repackage some of the results from §1.5. We assign to each rectangle in the plane two sequences, a *mass sequence* and a *capacity sequence*. We establish that these sequences determine the structure of the plaid model inside the rectangle.

In §25.5 we prove a technical result about vertical light points.

In §25.6 we give a criterion for the equation $PL' \cap S' = PL \cap S'$. Here PL is the set of plaid polygons with respect to p/q. Since the translation Υ is the identity, we can ignore it here. In §25.7 we verify the conditions of the Matching Criterion.

25.2 THE FIRST TWO STATEMENTS

We use the notation from the Copy Lemma, so that $(p'/q', p/q)$ is a pair of even rationals that appear in some core predecessor sequence. Here we establish part of the Copy Lemma in the weak and strong cases.

Lemma 25.1 $\Upsilon(S') \subset R^*$.

Proof: Here Υ is the identity. So, we just have to show that

- the top of S' is not above the top of R^*, and

- the right edge of S' does not lie to the right of the right edge of R^*.

For the right edge, it suffices to show that the width of S' is at most the width of R^*. The width of S' is $\min(\tau', \omega' - 2\tau')$. By Lemma 24.2, the width of R^* is τ. Statement 3 of Lemma 24.1, which applies because $\kappa = 0$ here, says that $\tau' + \tau = \omega'$, Since $\tau' < \omega'/2$, we get $\tau' \leq \tau$. This does it. The weak case has the same proof, except now Statement 3 tells us that $\tau' = \tau$.

Now consider the top edge for the strong case. It suffices to show that the height of S', namely ω', is not larger than the height of R^*, namely $\omega - \tau$. But $\omega' = \omega - 2\tau$, by Statement 2 of Lemma 24.1. This does it. The same argument works *a fortiori* in the weak case because the height of S' is less than ω'. $\square$

Here is the second statement of the Copy Lemma.

Lemma 25.2 $\Upsilon(H_2') \cap LH_2 \neq \emptyset$.

Proof: The weak and strong case have the same proof. Here Υ is the identity. The lines of H_2' are $y = \tau'$ and $y = \omega' - \tau'$. The line of LH_2 is $y = \tau$. We just have to prove that either $\tau' = \tau$ or $\omega' - \tau' = \tau$. But this is exactly Statement 3 of Lemma 24.1.

For the core case, we switch notation and consider the pair $(\overline{p}/\overline{q}, p/q)$. In this case we just have to show the spacing between the lines of $\overline{H}_2$ is the same as the spacing between the lines of H_2. The former quantity is $\overline{\omega} - 2\overline{\tau}$ and the latter quantity is $\omega - 2\tau$. But these two quantities are equal, by Statement 2 of Lemma 24.1. $\square$

We end with three more technical results that will be useful in the proof of the Copy Lemma. In these results, the slanting lines refer to the lines used in §1.5 to define the plaid model.

25.3 A TECHNICAL LEMMA

Lemma 25.3 *The slanting lines through the point $(0, \omega')$, relative to the parameter p/q, have capacity $\pm(\omega - 2)$.*

Proof: Relative to p/q, the mass of any slanting lines through $(0, \omega')$ is

$$(2p\omega' + \omega)_{2\omega}.$$

Since p'/q' is the even predecessor of p/q, we have

$$|2p\omega' - 2p'\omega| = 2.$$

From this we see that $2p\omega' + \omega$ differs from an odd multiple of ω by ± 2. Hence the capacity of any slanting line through $(0, \omega')$ is $\pm(\omega - 2)$. $\square$

25.4 THE MASS AND CAPACITY SEQUENCES

Let S' be as in the Copy Lemma. We fix integers $1 = x_0 < x_1 < \omega$ and $y_0 < y_1$ such that

$$S' = [x_0, x_1] \times [y_0, y_1]. \qquad (25.1)$$

The vertical grid lines intersecting S' have nonzero signed capacity. We define two sequences $\{c_j\}$ and $\{m_j\}$ w.r.t. S'. We have the *capacity sequence*

$$c_j = [4pj]_{2\omega}, \qquad j = x_0, ..., x_1. \qquad (25.2)$$

These are the sequences of capacities for the vertical lines through $(x_j, 0)$ for the given indices.

We also have the *mass sequence*

$$m_j = [2pj + \omega]_{2\omega}, \qquad j = y_0 - 2x + 1, ..., y_1 + 2x - 1, \qquad x = x_0 - x_1. \qquad (25.3)$$

These are the sequence of masses of the slanting lines through $(0, y_j)$ for the given indices.

When $j \in \omega' \mathbf{Z}$ the expression m'_j is either ± 1. It does not have a well-defined sign. The same goes for m_j when $j \in \omega \mathbf{Z}$. We call such indices *tricky indices*. As the name indicates, these indices are tricky to deal with. We first prove a result that limits the number of tricky indices we have to consider.

Lemma 25.4 *The only tricky indices that appear in the mass sequence for* p'/q' *are* 0 *and* ω'. *The only tricky indices that appear in the mass sequence for* p/q *is* 0.

Proof: First we consider p'/q'. In all cases, the smallest is $-2\mu' + 1$ where $\mu' = \min(\tau', 2\omega' - \tau')$. But $\mu < \omega'/2$. So, the smallest tricky index is larger than ω'. The largest tricky index is $\omega' + 2\mu' - 1 < 2\omega'$.

Now we consider p/q. Since $\omega' < \omega$ we just have to show that the index ω does not occur. This amounts to showing that $\omega' + 2\mu' \leq \omega$. But

$$\omega' + 2\mu' \leq \omega' + 2\tau' \leq \omega' + 2\tau = \omega.$$

The last equation is Statement 2 of Lemma 24.1. We have also used the fact that $\tau' \leq \tau$, which has the same proof as in Lemma 25.1. $\square$

Lemma 25.5 *Suppose that we know how the plaid polygons intersect a single horizontal grid line inside* S'. *Then the intersection of the plaid polygons with* S' *is determined by the signs of the mass and capacity sequences.*

Proof: Given our second definition of the plaid model in §1.5, the shade (light versus dark) of any vertical intersection point in S' is determined by the signs of the terms in the mass and capacity sequences.

We will suppose that we know how the plaid polygons intersect the bottom edge of S'. The case for any other edge has a similar treatment. Let Q be some unit square in S' for which we have not yet determined the plaid model inside Q. We can take Q to be as low as possible. But then we know how the plaid model intersects the bottom edge of Q, and the signs of the mass and capacity sequences determine how the tiling intersects the left and right edges. But we know that the boundary of any unit integer square intersects the plaid polygons in either 0 or 2 points. This allows us to determine how the plaid polygons intersect the top edge of Q. $\square$

25.5 VERTICAL INTERSECTION POINTS

Here we mention a fact about vertical light points that will be useful below. Let $m(\cdot)$ and $c(\cdot)$ respectively denote signed mass and capacity. Let $\sigma(\cdot)$ denote the sign of a line in the plaid model.

Lemma 25.6 *Suppose ζ is an intersection point of the form $V \cap L_1 \cap L_2$, where L_1 is a slanting line of negative slope and L_2 is a slanting line of positive slope and V is a vertical grid line. Then*

1. *If ζ is a light point, then $\sigma(L_2) = -\sigma(V)$.*

2. *If ζ is a light point then $|m(L_1)| + |m(L_2)| = |c(V)|$.*

3. *If ζ is a dark point and $m(L_2) < c(V)$ then $\sigma(L_2) = \sigma(V)$.*

Proof: Statement 1 follows immediately from the second definition of the plaid model. For Statement 2, let p/q be the parameter. Let x be the X-intercept of V and let y_j be the Y-intercept of L_j. We have

$$m(L_1) = (2py_1 + \omega)_{2\omega}, \qquad m(L_2) = (2py_2 + \omega)_{2\omega}, \qquad c(V) = (4px)_{2\omega}.$$

Since $P + Q = 2$, we have

$$\mathrm{slope}(L_2) - \mathrm{slope}(L_1) = 2.$$

Hence

$$y_1 - y_2 = 2x. \tag{25.4}$$

But then

$$m(L_1) - m(L_2) \equiv c(V) \bmod 2\omega. \tag{25.5}$$

Since $m(L_1)$ and $m(L_2)$ have opposite signs, we have

$$|m(L_1)| + |m(L_2)| \equiv |c(V)| \bmod 2\omega.$$

But $|c(V)| < \omega$. So this last equation remains true simply as an equation between positive integers.

For Statement 3, we let ρ denote reflection in the X-axis. Then $\rho(\zeta)$ is again a dark point and $L_2' = \rho(L_2)$ is a slanting line of negative slope having the same mass and opposite sign as L_2. If $\sigma(L_2) = -\sigma(V)$ then $\sigma(L_2') = \sigma(V)$. Also, we have $|m(L_2')| = |m(L_2)| < |m(V)|$. But then $\rho(\zeta)$ is a light point, by definition. But ρ permutes the light points. This is a contradiction. $\square$

25.6 A MATCHING CRITERION

We will give a criterion which guarantees that

$$PL \cap S' = PL' \cap S'. \tag{25.6}$$

Arithmetic Alignment: We say that the triple (S', PL, PL') is *arithmetically aligned* if

$$\text{sign}(c_j') = \text{sign}(c_j), \quad \forall j, \qquad \text{sign}(m_j') = \text{sign}(m_j), \quad \forall j \notin \{0, \omega'\}. \tag{25.7}$$

We are excluding these tricky indices in the second equation because at least one of the signs is not defined.

Geometric Alignment: For every pair of indices $\{i, j\}$ there are vertical intersection points z and z' so that z lies on a slanting line of slope $-P$ through $(0, i)$ and a slanting line of slope Q through $(0, j)$. We write $z \leftrightarrow z'$ where z' is the intersection of the slanting line of slope $-P'$ through $(0, i)$ and the slanting line of slope Q' through $(0, j)$. We also write $z \leftrightarrow z'$ for the similar relation that holds for the slanting lines of slope $(P, -Q)$ and those of slope $(P', -Q')$.

When $z \leftrightarrow z'$, we say that z and z' are *geometrically aligned* if these points are contained in the same unit vertical segment of S'. We call (S', PL, PL') *geometrically aligned* if $z \leftrightarrow z'$ are geometrically aligned whenever one of z or z' lies in S'. If this happens, then both points lie in S'.

Now we come to our Matching Criterion.

Lemma 25.7 (Matching Criterion) *Let S', PL, and PL' be as in the weak or strong case of the Copy Lemma. Suppose that (S', PL, PL') is arithmetically aligned and geometrically aligned. Then Equation 25.6 holds. That is, the Copy Lemma is true for the corresponding pair of parameters.*

Proof: Note that $PL \cap S'$ and $PL' \cap S'$ agree along the bottom grid line of S' because this line is part of the edge of the 0th block. Given Lemma 25.5, the present lemma is practically a tautology. The procedure given in Lemma 25.5 assigns exactly the same tiles to $S' \cap PL$ as it does to $S' \cap PL'$, provided that the right things happen with the tricky indices.

Now we will analyze what can go wrong with the tricky indices. We use the notation from Lemma 25.6. Also, we let $m'(\cdot)$ etc. denote the corresponding quantities relative to the parameter p'/q'. Before we get into the tedious details of this analysis, we explain the basic idea. There are not enough tricky indices to prevent us from using a second slanting line, though a non-tricky index, to decide whether a given intersection point is light or dark.

Now for the details. By Lemma 25.4, we just have to worry about 0 and ω'. The slanting lines through 0 have undefined signs for both parameters, so there are no light points on these lines. The only thing that can go wrong is if one of the slanting lines through $(0, \omega')$ has a vertical light point $\zeta \in S'$. We will assume that ζ exists and derive a contradiction.

Let L_1 be the slanting line through $(0, \omega')$ which contains ζ. In all cases, $(0, \omega')$ lies either above the top of S' or on the top edge of S'. In either case, L_1 has negative slope. Let V be the vertical line containing ζ. Since ζ is a light point, $|m(L_1)| < |c(V)|$. By Lemma 25.3, we have $|m(L_1)| = \omega - 2$. Hence $|c(V)| = \omega - 1$. There is a second line L_2, having the opposite type and positive slope, so that $\zeta \in L_1 \cap L_2 \cap V$. By Lemma 25.6, we have $|m(L_2)| = 1$. Hence L_2 has Y-intercept either τ or $\omega - \tau$. By Statement 2 of Lemma 24.1 we have $\omega - \tau > \omega'$. Since the Y-intercept of L_2 lies below the Y-intercept of L_1, the Y-intercept of L_2 is $(0, \tau)$.

By Statement 3 of Lemma 24.1, we have $\tau = \tau'$ or $\tau = \omega - \tau'$. In either case, $|m'(L_2)| = 1$. Note that $|m'(L_2)| < |m(V)|$ because $|m'(L_2)|$ is as small as possible. Since the counterpart of ζ is not a light point with respect to p'/q', and $|m'(L_2)| < |c'(V)|$, Lemma 25.6 says that $\sigma'(L_2) = \sigma'(V)$. But this contradicts the fact that $\sigma(L_2) = -\sigma(V)$. $\square$

25.7 VERIFYING THE MATCHING CRITERION

Lemma 25.8 (S', PL, PL') *is geometrically aligned.*

Proof: Suppose $z \leftrightarrow z'$. Note that z and z' lie on the same vertical line because $P + Q = P' + Q'$. The distance from this vertical line to the Y-axis is at most τ'. The difference in the slopes of the two relevant lines of the same type is

$$|P - P'| = |Q - Q'| = \frac{2}{\omega\omega'}.$$

Hence

$$\|z - z'\| \leq \frac{2\tau'}{\omega\omega'} < \frac{1}{\omega} < \frac{1}{\omega'}. \tag{25.8}$$

But z' is at least $1/\omega'$ from the interval containing it. Hence z and z' lie in the same vertical unit interval. $\square$

Lemma 25.9 *The two capacity sequences agree.*

Proof: This proof only has to do with the width of S', which is at most τ'. Define

$$C_j = 4pj, \qquad C'_j = 4p'j, \qquad \lambda = \omega'/\omega < 1. \qquad (25.9)$$

We have

$$c_j = [C_j]_{2\omega}, \qquad c'_j = [C'_j]_{2\omega'}, \qquad \lambda c_j = [\lambda C_j]_{2\omega'}.$$

We compute

$$|C'_j - \lambda C_j| \le 2j\omega'|P - P'| = \frac{2j}{\omega} \le \frac{2j\lambda}{\omega'} \le \frac{2\tau'\lambda}{\omega'} \le \lambda < 1.$$

Since $[C'_j]_{2\omega'}$ is always a nonzero even integer, it is at least 2 units from 0 and at least 1 unit from $\pm\omega$. But then λC_j and C'_j lie in the same interval $(k\omega', (k+1)\omega')$ for some $k \in \mathbf{Z}$. Hence c_j and c'_j have the same sign for all j. □

Now we deal with the mass sequences. There are 4 cases. We will consider these cases one at a time. We do most of the work in the first case, and then just describe the differences for the remaining cases.

Lemma 25.10 *The mass sequences agree in the weak case when $\tau' \le \omega' - 2\tau'$.*

Proof: We proceed as in the previous result. Define

$$M_j = 2pj + \omega, \qquad M'_j = 2p'j + \omega', \qquad \lambda = \omega'/\omega < 1. \qquad (25.10)$$

We have

$$m_j = [M_j]_{2\omega}, \qquad m'_j = [M'_j]_{2\omega'}, \qquad \lambda m_j = [\lambda M_j]_{2\omega'}. \qquad (25.11)$$

We show that m'_j and λm_j have the same sign for all relevant j.

Let W' and H' respectively denote the width and height of S'. We have $|j| < H' + 2W'$. Hence

$$\Delta_j := |M'_j - \lambda M_j| = \frac{2|j|}{\omega} < \frac{2H' + 4W'}{\omega}. \qquad (25.12)$$

Statements 2 and 3 of Lemma 24.1 tell us that

$$\tau' = \tau, \qquad \omega' = \omega - 2\tau.$$

Also, in this case, $W' = \tau'$ and $H' = \omega' - \tau'$. Hence

$$\Delta_j < \frac{2\omega' + 2\tau'}{\omega} = \frac{2\omega' + 2\tau}{\omega} = \frac{2\omega - 2\tau}{\omega} = \frac{\omega + \omega'}{\omega} = 1 + \lambda.$$

Since m'_j and ω' are both odd, there are 3 cases:

- Suppose $1 < |m'_j| < \omega' - 1$. Here M'_j is at least 2 units from the endpoints of the interval $I_k(k\omega', (k+1)\omega')$ in which it lies. Hence λM_j lies in the same interval. We call I_k an ω-*interval*.

- Suppose $|m'_j| = 1$. If λM_j does not lie in the same ω'-interval as M'_j then we have $|\lambda M_j - k\omega'| < \lambda$ for some integer k. Dividing through by λ we get $|M_j - k\omega| < 1$. Since all the terms are integers, we must have $M_j = k\omega$. This is only possible if ω divides j. But Lemma 25.4 rules this out.

- If $|m'_j| = \omega'$ then ω' divides j. But then j is a tricky index and we don't need to consider it.

This completes the proof. □

Lemma 25.11 *The mass sequences agree in the weak case when $\tau' > \omega' - 2\tau'$.*

Proof: The proof is just like this previous case except that now we have $W' = \omega' - 2\tau'$ and $H' = 2\tau'$. Plugging this into Equation 25.12 gives

$$\Delta_j < \frac{\omega' + 4\tau'}{\omega} = \frac{\omega + 2\tau'}{\omega} \leq \frac{\omega + \omega'}{\omega} = 1 + \lambda. \qquad (25.13)$$

This is the same as in the previous case. □

Lemma 25.12 *The mass sequences agree in the strong case if $\tau' \leq \omega' - 2\tau'$.*

Proof: This time $\tau + \tau' = \omega'$. Hence $2\tau' \leq \omega' - \tau' = \tau$. This time we have $W' = \tau'$ and $H' = \omega'$. Plugging this in to Equation 25.12 gives

$$\Delta_j < \frac{2\omega' + 4\tau'}{\omega} \leq \frac{2\omega' + 2\tau}{\omega} = \frac{(\omega' + 2\tau) + \omega'}{\omega} = \frac{\omega + \omega'}{\omega} = 1 + \lambda. \quad (25.14)$$

This is the same as in the previous cases. □

Lemma 25.13 *The mass sequences agree in the strong case if $\tau' > \omega' - 2\tau'$.*

Proof: In this case, we have $H' = \omega'$ and $W' = \omega' - 2\tau'$. Plugging this in to Equation 25.12 gives

$$\Delta_i < \frac{6\omega' - 8\tau'}{\omega}. \qquad (25.15)$$

This step requires some algebraic manipulation. Using the fact that $\tau' = \omega' - \tau$ and $\omega' + 2\tau = \omega$, we have

$$6\omega' - 8\tau' = 6\omega' - 4\tau' - 4\tau' = 6\omega' - 4(\omega' - \tau) - 4\tau' =$$

$$2\omega' + 4\tau - 4\tau' = (\omega' + 2\tau) + (\omega' + 2\tau) - 4\tau' = \omega + \omega' + (2\tau - 4\tau').$$

Since $\tau' > \omega'/3$ and $\tau + \tau' = \omega'$ we get $\tau < 2\omega/3$. Hence $2\tau - 4\tau' < 0$. Our big calculation therefore shows that

$$6\omega' - 8\tau' < \omega + \omega'.$$

Hence $\Delta_j < 1 + \lambda$ as in the previous cases. □

Chapter Twenty-Six

The Core Case

26.1 CHAPTER OVERVIEW

In this chapter we prove the core case of the Copy Lemma. The proof follows the same strategy as in the previous chapter, but it is considerably harder. I think that the difficulty might be necessary because the result is quite sharp. We will put off the difficult steps as long as possible, but in the end we will have to face them. We switch notation so that our pair is $(\overline{p}/\overline{q}, p/q)$.

In §26.2 we prove the first two statements of the Copy Lemma. The rest of the chapter is devoted to the third statement.

In §26.3 we define geometric and arithmetic alignment as in the previous chapter, but with the twist that there are some extra indices we have to look after carefully.

In §26.4 we verify the geometric alignment criterion just as in the previous chapter.

In §26.5 we show that the signs of the two capacity sequences match. This is half of the arithmetic alignment.

In §26.6 we present a technical lemma which will help us deal with the mass sequences.

For the mass sequences we divide the indices into two kinds: *central* and *peripheral*. We deal with the central indices with an estimate like that in the previous chapter. In §26.7 we show that the signs of the mass sequences agree on the central indices. In §26.8 we verify that the signs of the mass sequences agree on the peripheral sequences except for two special indices where the signs can (and sometimes do) disagree. This one mismatch turns out not to be a problem.

In §26.9 we finish the proof of this long and difficult argument.

The Basic Facts: Since it comes up often, we recall that

- $3\tau < \omega$.

- $3\overline{\tau} < \overline{\omega}$.

- $\omega - 2\tau = \overline{\omega} - 2\overline{\tau}$.

The first two statements come from Lemma 24.2 and the last one is Statement 2 of Lemma 24.1. We call these the *basic facts*.

26.2 THE FIRST TWO STATEMENTS

We use the notation from the Copy Lemma. Here is the first statement of the Copy Lemma. We switch notation to reflect that we are using $\overline{p}/\overline{q}$ in place of p'/q'.

Lemma 26.1 $\Upsilon(\overline{S}) \subset R^*$.

Proof: As in the proof of Lemma 25.1, we just have to consider what Υ does to the top edge and the right edge. Consider the right edge. The width of $\overline{S}$ is $\overline{\tau}$ and the width of R^* is $\omega - 2\tau$. The latter quantity is larger, by the basic facts. Now consider the top edge. The distance from the horizontal midline of the block $\overline{B}$ to the top of $\overline{S}$ is

$$\overline{d} = \overline{\omega} - \overline{\tau} - \overline{\omega}/2 = \overline{\omega}/2 - \overline{\tau} = \omega/2 - \tau.$$

The last equality is a basic fact. The distance from the horizontal midline of the block B to the top of R^* is

$$d = \omega - (\omega - 2\tau) - \omega/2 = 2\tau - \omega/2.$$

The inequality $\overline{d} \leq d$ is equivalent to a basic fact. □

Here is the second statement of the Copy Lemma.

Lemma 26.2 $\Upsilon(\overline{H}_2) = H_2$.

Proof: In this case we just have to show the spacing between the lines of $\overline{H}_2$ is the same as the spacing between the lines of H_2. The former quantity is $\overline{\omega} - 2\overline{\tau}$ and the latter quantity is $\omega - 2\tau$. These are equal, by a basic fact. □

26.3 GEOMETRIC AND ARITHMETIC ALIGNMENT

We have

$$\overline{S} = [0, \overline{\tau}] \times [0, \overline{\omega}], \qquad \Upsilon(\overline{S}) = [0, \overline{\tau}] \times [\omega/2 - \overline{\omega}/2, \omega/2 + \overline{\omega}/2]. \quad (26.1)$$

The capacity sequence associated to p/q is

$$c_j = [4pj]_{2\omega}, \qquad j = 1, ..., \overline{\tau}. \quad (26.2)$$

The capacity sequence associated to $\overline{p}/\overline{q}$ has the same form. Because the translation Υ is nontrivial in this case, we have to take a little more care in defining the mass sequences. The mass sequence associated to p/q is

$$m_j, \qquad j = -a + 1, ..., a - 1, \qquad a = (\omega/2 - \overline{\omega}/2) - 2\overline{\tau}. \quad (26.3)$$

The mass sequence associated to $\overline{p}/\overline{q}$ is

$$\overline{m}_j, \qquad j = -2\overline{\tau} + 1, ..., \overline{\omega} + 2\overline{\tau} - 1. \quad (26.4)$$

If is convenient to define

$$\bar{j} = j - (\omega/2 - \overline{\omega}/2). \tag{26.5}$$

By construction, $\Upsilon(0, \bar{j}) = (0, j)$.

Arithmetic Alignment: We say that $(\overline{S}, PL, \overline{PL})$ is *arithmetically aligned* if

$$\text{sign}(\overline{c}_j) = \text{sign}(c_j), \quad \forall j,$$

$$\text{sign}(\overline{m}_{\bar{j}}) = \text{sign}(m_j), \quad \forall j \neq \omega/2 \pm \overline{\omega}/2, \omega/2 \pm \omega/2 \pm \eta. \tag{26.6}$$

Here $\eta = \omega - 2\tau$. For these indices, we have either $|\overline{m}_{\bar{j}}| = \overline{\omega}$ or $\overline{m}_{\bar{j}} = \pm(\overline{\omega} - 2)$. In the former case, we are just excluding *tricky indices* as in the previous chapter. In the latter case, we genuinely can have a sign mismatch. We will call the corresponding indices *super tricky* to indicate their role played in the proof.

Geometric Alignment: The geometric alignment works the same way as in the previous chapter, except now we match up a point $\overline{z}$ in some vertical unit interval $\overline{I}$ with the corresponding point in the interval $\Upsilon(\overline{I})$.

26.4 GEOMETRIC ALIGNMENT

Here we prove that $(\overline{S}, PL, \overline{PL})$ is geometrically aligned. Let $\overline{\zeta}$ and ζ be corresponding vertical intersection points. Let $\overline{I}$ and I be the two vertical intervals respectively containing $\overline{\zeta}$ and ζ. We need to prove that $\Upsilon(\overline{I}) = I$.

Let $\overline{L}$ and L be two slanting lines of the same type which contain $\overline{\zeta}$ and ζ respectively. The two lines $\Upsilon(\overline{L})$ and L have the same Y-intercept. The difference in their slopes is

$$|P - \overline{P}| = |Q - \overline{Q}| = \frac{4\kappa}{\omega\overline{\omega}}. \tag{26.7}$$

Here $\kappa \geq 1$. These lines move at most $\overline{\tau}$ in the horizontal direction. Hence

$$\|\zeta - \Upsilon(\overline{\zeta})\| \leq \frac{4\kappa\overline{\tau}}{\omega\overline{\omega}}. \tag{26.8}$$

Using the basic facts and Statement 4 of Lemma 24.1, we have

$$\kappa\overline{\tau} < \frac{1}{3}\kappa\overline{\omega} < \frac{\kappa}{3 + 2\kappa}\omega < \frac{\omega}{2}.$$

Combining our equations, we get that

$$\|\zeta - \Upsilon(\overline{\zeta})\| < 2/\overline{\omega}. \tag{26.9}$$

We get the desired result as long as $\overline{\zeta}$ is at least $2/\overline{\omega}$ from the boundary $\partial\overline{I}$. We just have to worry about the case when $\overline{\zeta}$ is less than $2\overline{\omega}$ from $\partial\overline{I}$.

Lemma 26.3 *Let V be the vertical grid line in $\overline{S}$ that contains $\overline{I}$. If $\overline{\zeta}$ is less than $2/\overline{\omega}$ from $\partial\overline{I}$ then V is the vertical line through $(\overline{\tau})$.*

Proof: $\overline{\zeta}$ lies on a slanting line $\overline{L}$ of slope $\overline{P}$. Since $\overline{P} = 2\overline{p}/\overline{\omega}$, the distance from $\overline{\zeta}$ to $\partial \overline{I}$ is some integer multiple of $1/\overline{\omega}$. We just have to understand what happens when the distance is precisely $1/\overline{\omega}$.

Suppose that V has the equation $x = x_0$. Let y_0 be the Y-intercept of $\overline{L}$. The point $\overline{\zeta}$ lies at the point

$$\left(x_0, y_0 + \frac{2\overline{p}x_0}{\overline{\omega}} \right). \tag{26.10}$$

If $\overline{\zeta}$ is $1/\overline{\omega}$ from $\partial \overline{I}$ then we must have $2\overline{p}x_0 \equiv \pm 1 \bmod \overline{\omega}$. This forces $x_0 = \pm\overline{\tau}$. The latter case leads to a line outside $\overline{S}$. Hence $x_0 = \overline{\tau}$. □

We just have to deal with the case when our intersection point $\overline{\zeta}$ lies on the line $x = \overline{\tau}$. In this case, $\overline{\zeta}$ either lies $1/\overline{\omega}$ units above the lower endpoint of $\partial \overline{I}$ or $1/\overline{\omega}$ units below the upper endpoint. The choice depends on whether $2\overline{p}\overline{\tau}$ is congruent to 1 or $-1 \bmod \overline{\omega}$. Call this the *sign choice*. The worry is that when we switch from $\overline{p}/\overline{q}$ to p/q the sign choice switches, causing the corresponding point ζ to switch into the wrong interval. But Statement 6 of Lemma 24.1 guarantees that the two sign choices are the same.

This completes the proof of geometric alignment.

26.5 ALIGNMENT OF THE CAPACITY SEQUENCES

Our argument is like the proof of Lemma 25.9. Define

$$C_j = 4pj, \qquad \overline{C}_j = 4\overline{p}j, \qquad \lambda = \overline{\omega}/\omega < 1. \tag{26.11}$$

We have

$$c_j = [C_j]_{2\omega}, \qquad \overline{c}_j = [C_j]_{2\overline{\omega}}, \qquad \lambda c_j = [\lambda C_j]_{2\overline{\omega}}.$$

We compute

$$\Delta_j := |\overline{C}_j - \lambda C_j| \le 2j\overline{\omega}|P - \overline{P}| = \frac{8\kappa j}{\omega} \le \frac{8\kappa\overline{\tau}}{\omega} < \frac{8\kappa\overline{\omega}}{3\omega} < \frac{8\kappa}{3 + 2\kappa} < 4. \tag{26.12}$$

In this estimate, we used the basic fact, and Statement 4 of Lemma 24.1.

The argument in Lemma 25.9 works just as it did there except for the following cases:

1. $\overline{c}_j = \pm 2$.

2. $\overline{c}_j = \pm(\overline{\omega} - 1)$.

3. $\overline{c}_j = \pm(\overline{\omega} - 3)$.

Case 1: We will consider the case when $\overline{c}_j = 2$. The case when $\overline{c}_j = -2$ has the same treatment. If $\overline{c}_j = 2$ then $j = \overline{\tau}$ and the corresponding vertical grid line has the equation $x = \overline{\tau}$, as in the end of the proof in the previous section. Statement 7 of Lemma 24.1 now says that $c_{\overline{\tau}} = 4\kappa + 2$. The signs match in this case.

Case 2: We treat the case when $\bar{c}_j = \omega - 1$. By Lemma 1.1 we have either $j = \bar{\tau}/2$ (and $\bar{\tau}$ is even) or $j - \bar{\omega}/2 - \bar{\tau}/2$ (and $\bar{\tau}$ is odd). The latter case is irrelevant because the corresponding line is outside $\overline{S}$. By definition, we have

$$4\bar{p}j \equiv \bar{\omega} - 1 \bmod 2\bar{\omega}.$$

But then

$$4\bar{p}(2j) = 4\overline{p\tau} \equiv 2\omega - 2 \bmod 2\omega.$$

Hence $c_{2j} = -2$. But then $\bar{c}_{2j} = -(4\kappa + 2)$, by Case 1. Hence

$$4p(2j) \equiv 2\omega - (4\kappa + 2) \bmod 2\omega.$$

Hence

$$4pj \equiv \omega - (2\kappa + 1) \bmod 2\omega. \tag{26.13}$$

Since $\bar{\omega} = \omega - 2\kappa\omega'$, we see that $2\kappa + 1 < \omega$. Hence the quantity in Equation 26.13 is positive. Hence $\bar{c}_j$ and c_j have the same sign.

Case 3: We must have $8\kappa j/\omega \geq 3$ in order for our estimate not to rule out a sign mismatch. Since $j \leq \hat{\tau}$, we have

$$\frac{12\kappa\bar{\tau}}{\omega} > \frac{8\kappa\bar{\tau}}{\omega} \geq \frac{8\kappa j}{\omega} \Delta_j \geq 3. \tag{26.14}$$

When $\bar{\tau}$ is even, we have $j = 3\bar{\tau}/2$ or $j = \bar{\omega} - 3\bar{\tau}/2 > 3\bar{\tau}/2$. The last inequality uses the basic fact.

When $\bar{\tau}$ is odd, we have $j = \bar{\omega}/2 - 3\bar{\tau}/2$. We compute

$$\Delta_j = \frac{8\kappa j}{\omega} = \frac{4\kappa\bar{\omega}}{\omega} - \frac{12\kappa\bar{\tau}}{\omega} \leq_* 6 - \frac{12\kappa\bar{\tau}}{\omega} < 3. \tag{26.15}$$

The starred inequality follows from Statement 4 of Lemma 24.1. The last inequality comes from Lemma 26.14.

26.6 A TECHNICAL LEMMA

Let J denote the interval of length $\bar{\omega}$ centered at $\omega/2$. Let J_+ denote the interval of length $2\bar{\tau}$ whose left endpoint coincides with the right endpoint of J. Let J_- be the symmetrically placed interval on the other side of J. Each central index i corresponds to $j \in J$. Each peripheral index i corresponds to j in the interior of $J_+ \cup J_-$. We will assume in our result that $2p\tau \equiv +1 \bmod \omega$. Each of the statements has a symmetric formulation when $2p\tau \equiv -1 \bmod \omega$.

Lemma 26.4 *Suppose that $2p\tau \equiv 1 \bmod \omega$.*

1. *If j is the right endpoint of J then $m_j = -\omega + 2\kappa$.*

2. If $j \in J$ and $|m_j| \le 4\kappa + 1$ then $|m_j| = 1$.

3. There is no $j \in J_+$ with $m_j \in [+1, +(4\kappa - 1)]$.

4. There is no $j \in J_-$ with $m_j \in [-(4\kappa - 1), -1]$.

5. There is at most one j in the interior of J_+ with the property that $m_j \in [-\omega + 2, -\omega + (4\kappa - 2)]$. For this index we have $|\overline{m}_{\overline{j}}| = \omega - 2$.

6. There is at most one j in the interior of J_- with the property that $m_j \in [+\omega - 2, +\omega - (4\kappa - 2)]$. For this point we have $|\overline{m}_{\overline{j}}| = \omega - 2$.

Proof: Let $\eta = \omega - 2\tau = \overline{\omega} - 2\overline{\tau}$. When an index changes by $+\eta$ the signed mass changes by $+2 \bmod 2\omega$, both with respect to p/q and with respect to $\overline{p}/\overline{q}$. We will use this fact repeatedly in our arguments. For reference, we call it the *mass jump property*.

Statement 1: We have

$$m_j = (p\omega + p\overline{\omega} + \omega)_{2\omega}. \qquad (26.16)$$

We will compute everything mod 2ω. Using Statements 1 and 2 of Lemma 24.1, we compute

$$p\omega + p\overline{\omega} + \omega \equiv$$

$$-p\omega + p\overline{\omega} + \omega \equiv$$

$$-2p\kappa\omega' + \omega \equiv$$

$$-2p\kappa(\omega - 2\tau) + \omega \equiv$$

$$4p\kappa\tau + \omega \equiv$$

$$2\kappa(2p\tau) + \omega \equiv$$

$$2\kappa(1 + \omega) + \omega \equiv$$

$$2\kappa + \omega \equiv$$

$$-\omega + 2\kappa \bmod 2\omega.$$

But then the quantity in Equation 26.16 is $-\omega + 2\kappa$ provided that this quantity lies in $(-\omega, \omega)$. In fact, the quantity lies in $(-\omega, 0)$ because $2\kappa < \omega$. To see this note that $\overline{\omega} = \omega - 2\kappa\omega'$.

Statement 2: We will consider the indices j with $m_j = 3, 5, 7, ...$ The opposite case, when $m_j = -3, -5, -7, ...$, has the same treatment. Let $J_\#$ be interval of length $\omega - \overline{\omega}$ whose left endpoint coincides with the right endpoint of J. Notice that no point of $J_\#$ is equivalent to a point of J mod

ω. So, to prove our result, it suffices to show that $J_{\#}$ contains the indices j with $m_j = 3, 5, ..., 4\kappa + 1$.

Given that $m_{\omega-\tau} = -1$ and $m_\tau = 1$ and the mass jump property, the indices we need to consider are

$$j(\lambda) = \tau - \lambda\eta, \qquad \lambda = 1, 2, ..., 2\kappa. \qquad (26.17)$$

We just have to show that all these points belong to $J_{\#}$. The points occur in arithmetic progression, so it suffices to prove that $j(1)$ lies to the left of the right endpoint of $J_{\#}$ and $j(2\kappa)$ lies to the right of the left endpoint of $J_{\#}$. Multiplying through by 2, we get the following two inequalities that we need to check.

$$2\tau - 2\eta \leq \omega - \overline{\omega}, \qquad 2\tau - 4\kappa\eta \geq -\omega + \overline{\omega}. \qquad (26.18)$$

The first inequality is the same as

$$-2\omega + 6\tau \leq \omega - \overline{\omega}.$$

This follows from the basic facts. The second inequality is equivalent to

$$-4\kappa\omega + (8\kappa + 2)\tau \geq -\omega + \overline{\omega}.$$

By Statement 5 of Lemma 24.1, we have

$$\overline{\omega} = (1 - 2\kappa)\omega + 4\kappa\tau. \qquad (26.19)$$

Thus, our second inequality is equivalent to

$$-4\kappa\omega + (8\kappa + 2)\tau \geq -2\kappa\omega + 4\kappa\tau.$$

This is in turn equivalent to the statement that

$$\frac{\tau}{\omega} \geq \frac{2\kappa}{4\kappa + 2} = \frac{\kappa}{2\kappa + 1}.$$

This last inequality is true by the definition of κ given in Equation 24.1.

Statement 3: We are considering the same indices as in Statement 2. We just need to show that $j(\lambda)$ lies at least $2\hat{\tau}$ units to the right of the left endpoint of $J_{\#}$. But this follows from the fact that $j(2\kappa)$ lies to the right of the left endpoint of $J_{\#}$ and that $\eta < 2\hat{\tau}$. The last inequality is a consequence of the basic facts.

Statement 4: This has the same proof as Statement 3.

Statement 5: By Statement 1, the indices under discussion have the form

$$j(\lambda) = (\omega/2 + \overline{\omega}/2) + \lambda\eta, \qquad \lambda = -\kappa + 1, ..., \kappa - 1. \qquad (26.20)$$

The point $j(1)$ might lie in J_+ but note that $j(2)$ already lies to the right of J_+. This follows from the inequality

$$2\eta = 2\omega - 4\tau = 2\overline{\omega} - 4\overline{\tau} > 2\overline{\tau},$$

which follows from the basic facts.

For $\lambda = 0, ..., \kappa - 1$, we claim that the remaining indices all lie in the interval of length ω whose left endpoint coincides with the right endpoint of J. At the same time, for $\lambda = 0, ..., -\kappa + 1$, the corresponding indices all lie in the interval of length $\omega - 2\overline{\tau}$ whose right endpoint coincides with the right endpoint of J. Neither of these intervals has any points not already in J_+ that are equivalent mod ω to a point in the interior of J_+.

Our claim follows from the inequality

$$(\kappa - 1)\eta < \omega - 2\overline{\tau}.$$

The analysis in the proof of Statement 2 gives the inequality

$$(2\kappa - 2)\eta < \omega - \overline{\omega},$$

and this is stronger than the inequality we are seeking to prove here.

It just remains to compute m_j when $j = j(1)$. When $j = j_0$ we have $\overline{j} = \overline{\omega}$ and $|\overline{m_{\overline{j}}}| = \overline{\omega}$. The fact that $|\overline{m_{\overline{j}}}| = \overline{\omega} - 2$ now follows from the mass jump property.

Statement 6: This has the same kind of proof as Statement 5. $\square$

26.7 THE MASS SEQUENCES: CENTRAL CASE

Recall that

$$\overline{j} = j - (\omega/2 - \overline{\omega}/2). \qquad (26.21)$$

We want to show that m_j and $\overline{m_{\overline{j}}}$ have the same sign for all j except for the tricky indices $j = \omega/2 \pm \overline{\omega}/2$, which we can ignore. It is convenient to introduce the new indices

$$i = j - \omega/2 = \overline{j} - \overline{\omega}/2. \qquad (26.22)$$

Note that i is a half-integer. This does not bother us.

$$\mu_i = m_j, \qquad \overline{\mu}_i = \overline{m_{\overline{j}}}. \qquad (26.23)$$

We want to show that $\overline{\mu}_i$ and μ_i have the same sign for all relevant i. By symmetry, it suffices to take $i > 0$. The range is

$$i \in [1/2, \overline{\omega}/2 + 2\overline{\tau} - 1]. \qquad (26.24)$$

Now we proceed somewhat as in §25.11. Define

$$M_j = 2pj + \omega - p\omega, \qquad \overline{M}_j = \overline{p}j + \overline{\omega} - 2\overline{p\omega}, \qquad \lambda = \overline{\omega}/\omega < 1. \quad (26.25)$$

The last terms we have added require some explanation. By Statement 1 of Lemma 24.1, the integers p and $\overline{p}$ have the same parity. So, when these integers are even, the additional terms have no effect on the congruence of M_j and $\overline{M}_j$ mod 2ω and $2\overline{\omega}$. When the integers are odd, they both globally shift the values by ω, which amounts to switching the signs of m_j and $\overline{m}_{\overline{j}}$. Thus we have

$$\pm \mu_i = [M_j]_{2\omega}, \qquad \pm \overline{\mu}_i = [M_{\overline{j}}]_{2\overline{\omega}}, \qquad \pm \lambda \mu_i = [\lambda M_j]_{2\overline{\omega}}, \qquad (26.26)$$

and the sign is the same in all three equations.

Using the fact that $P\omega = 2p$ and $\overline{P}\overline{\omega} = 2\overline{p}$, we compute

$$\Delta_i := |\overline{M}_{\overline{j}} - \lambda M_j| = \overline{\omega}|(\overline{P}\overline{j} + 1 - \overline{p}) - (Pj + 1 - p)| =$$

$$\overline{\omega}|\overline{P}\overline{j} - Pj - \overline{p} + p| = \overline{\omega}|\overline{P}(i + \overline{\omega}/2) - P(i + \omega/2) - \overline{p} + p| =$$

$$\overline{\omega}|(\overline{P}i + \overline{p}) - (Pi - p) + \overline{p} - p| = |i||\overline{\omega}||\overline{P} - P| = \frac{4\kappa|i|}{\omega}.$$

In short,

$$\Delta_i = \frac{4\kappa|i|}{\omega}. \qquad (26.27)$$

We say that the index i is *central* if $i \in (-\overline{\omega}/2, \overline{\omega}/2)$. Otherwise, we call i *peripheral*.

Lemma 26.5 *If i is a central index then $\overline{\mu}_i$ and μ_i have the same sign.*

Proof: When i is a central index, we have

$$\Delta_i < \frac{2\kappa\overline{\omega}}{\omega} \le \frac{6\kappa}{3 + 2\kappa} < 3. \qquad (26.28)$$

The last inequality uses Statement 4 of Lemma 24.1. Just as we argued above, we see that μ_i and $\overline{\mu}_i$ have the same sign in all cases except when $\overline{\mu}_i = \pm 1$ or $\overline{\mu}_i = \pm(\overline{\omega} - 2)$. We will treat these cases in turn.

Case 1: If $\overline{\mu}_i = \pm 1$ then $\overline{j} = \overline{\tau}$ or $\overline{j} = \overline{\omega} - \overline{\tau}$. In this case, $j = \tau$ or $j = \omega - \tau$ by Lemma 25.2. But for these indices we have $\overline{m}_{\overline{j}} = \pm 1$ and $m_j = \pm 1$. The signs match for the same reason as considered in Case 1 of the previous section, namely Statement 7 of Lemma 24.1.

Case 2: Say that an $\overline{\omega}$-interval is an interval of the form $[k\overline{\omega}, (k + 1)\overline{\omega}]$ for some integer k. Likewise define an ω-interval. For the relevent indices j, the point $\overline{M}_{\overline{j}}$ lies 2 units from the endpoint of an $\overline{\omega}$-interval. Given our bound on Δ_i, this means that the corresponding point λM_j lies within 1 unit from the endpoint of an $\overline{\omega}$-interval. Scaling by λ^{-1} and using Statement 4 of Lemma 24.1, we see that M_j lies within $2\kappa + 1$ from the endpoint of an ω-interval. So, if μ_i and $\overline{\mu}_i$ do not have the same sign we have $|\mu_i| < 2\kappa + 1$. Statement 2 of Lemma 26.4 now says that $|\mu_i| = 1$. This contradicts the fact that the index i is not one of the indices considered in Case 1. $\square$

Remark: When we include the peripheral indices, we get the weaker bound $\Delta_i < 7$. This leaves too many special cases for us.

26.8 THE MASS SEQUENCES: PERIPHERAL CASE

We use the same intervals that we considered in Lemma 26.4. Let I, I_-, I_+ be the intervals obtained by translating J, J_-, J_+ by $-(\omega/2 - \overline{\omega}/2)$. Each $i \in I_*$ corresponds to $j \in J_*$. Thus I contains the central indices and the interior of $I_+ \cup I_-$ contains the peripheral indices.

Lemma 26.6 *If $i \in I^+$ then μ_i and $\overline{\mu}_i$ have the same sign unless $|\overline{\mu}_i| = \omega - 2$.*

Proof: Let $i' = i - \overline{\omega}$. We have $i' \in I$. Note that $\overline{\mu}_i = \overline{\mu}_{i'}$ because the mass sequence for $\overline{p}/\overline{q}$ is periodic with periodic $\overline{\omega}$. So, we just have to see that μ_i and $\mu_{i'}$ have the same sign except in the one case.

By Statement 1 of Lemma 26.4,

$$m_{\omega/2+\overline{\omega}/2} = \pm(\omega - 2\kappa).$$

We will consider the case when

$$m_{\omega/2+\overline{\omega}/2} = -\omega + 2\kappa. \qquad (26.29)$$

The other case has the same treatment. By symmetry,

$$m_{\omega/2-\overline{\omega}/2} = +\omega - 2\kappa. \qquad (26.30)$$

Given the affine nature of the dependence of μ on the indices, we have

$$\mu_{i'} = \mu_i - 4\kappa \bmod 2\omega. \qquad (26.31)$$

A direct calculation would also bear this out.

There are two ways it can happen that μ_i and $\mu_{i'}$ have the opposite signs:

1. $\mu_i \in [+1, +(4\kappa - 1)]$.

2. $\mu_i \in [-\omega + 2, -\omega + (4\kappa - 2)]$.

Statement 3 of Lemma 26.4 rules out Case 1. Statement 5 of Lemma 26.4 shows that Case 2 just has one exception, the one advertised in the lemma. $\square$

Lemma 26.7 *If $i \in I^-$ then μ_i and $\overline{\mu}_i$ have the same sign unless $|\overline{\mu}_i| = \omega - 2$.*

Proof: Let $i' = i + \overline{\omega}$. This time we have

$$\mu_{i'} = \mu_i + 4\kappa \bmod 2\omega. \qquad (26.32)$$

There are two ways it can happen that μ_i and $\mu_{i'}$ have the opposite signs:

1. $\mu_i \in [-(4\kappa - 1), -1]$.

2. $\mu_i \in [+\omega - (4\kappa - 2), +\omega - 2]$.

Statement 4 of Lemma 26.4 rules out Case 1. Statement 6 of Lemma 26.4 shows that Case 2 just has one exception, the one advertised in the lemma. $\square$

26.9 THE END OF THE PROOF

The following result completes the proof of the Copy Lemma.

Lemma 26.8 (Matching Criterion) *Let $\overline{S}$, PL, and $\overline{PL}$ be as in the core case of the Copy Lemma. Suppose that $(\overline{S}, PL, \overline{PL})$ is arithmetically aligned and geometrically aligned. Then the Copy Lemma is true for the corresponding pair of parameters.*

Proof: The proof is like the proof of Lemma 25.7 but harder. There are three main points.

1. We have to see that $\Upsilon(\overline{PL} \cap \overline{S})$ and $PL \cap \Upsilon(\overline{S})$ match along some horizontal grid line.

2. We have to deal with the tricky indices.

3. We have to deal with the super tricky indices.

Matching along a horizontal line: Let L be the horizontal line through $(0, \tau)$. As in the proof of Theorem 3.1, the line L only intersects the plaid polygons twice inside the block $[0, \omega]^2$. One of the intersections is the point on L having X-coordinate $1/2$. The other point lies to the right of the vertical midline of the block and hence is outside $\Upsilon(\overline{S})$. At the same time, the corresponding line $\overline{L}$ intersects $\overline{PL}$ in two points, one of them having X-coordinate $1/2$ and the other one outside $\overline{S}$. Finally, $\Upsilon(\overline{L} \cap \overline{S}) = L \cap \Upsilon(\overline{S})$. This gives us the matching along a horizontal line.

Tricky indices: Now we will analyze what can go wrong with the tricky indices. We just have to see that the slanting lines through the points $(0, \omega/2 \pm \overline{\omega}/2)$ have no light points, with respect to p/q, which lie in $\Upsilon(\overline{S})$. Suppose that ζ is such a light point. Let L_1 be the slanting line which contains ζ and one of our tricky points. Let L_2 be the other slanting line through ζ. Let V be the vertical line through ζ. We use the same notation for masses and capacities as in the proof of Lemma 25.7.

By Statement 1 of Lemma 26.4, we have $|m(L_1)| = \omega - 2\kappa$. Hence, by Lemma 25.6, we have $|m(L_2)| < 2\kappa$. Let y_j denote the Y-intercept of L_j. We have

$$|y_1 - y_2| \leq 2\overline{\tau} < \overline{\omega}.$$

This forces the line L_2 to have Y-intercept $(0, j)$ with $j \in J$. But then $|m(L_2)| = 1$ by Statement 2 of Lemma 26.4. This forces $y_2 = \omega - \tau$ or $y_2 = \tau$. But then the corresponding line $\overline{L}_2$ has Y-intercept $\overline{\tau}$ or $\overline{\omega} - \overline{\tau}$. But then $|m(L_2)| = 1$. Now we get the same contradiction as in the proof of Lemma 25.7.

Super tricky indices: Let j be a super tricky index. By Statements 5 and 6 of Lemma 26.4, we have

$$|m_j| > \omega - 4\kappa, \qquad |\overline{m}_{\overline{j}}| = \overline{\omega} - 2. \qquad (26.33)$$

We finish the proof by showing that none of the light points associated to these indices lies in the relevant rectangle.

Consider the parameter $\overline{p}/\overline{q}$ first. The index $\overline{j}$ corresponding to a super tricky index is either $-\eta$ and $\overline{\omega} + \overline{\eta}$. Let $\overline{L}_1$ be a slanting line through $(0, \overline{j})$ which intersects $\overline{S}$. Let $\overline{L}_2$ be the slanting line of the opposite slope such that $\overline{L}_1 \cap \overline{L}_2$ is the offending light point. By Lemma 25.6 we have $|m(\overline{L}_2)| = 1$. Hence $\overline{L}_2$ contains one of the two points $(0, \overline{\tau})$ and $(0, \overline{\omega} - \overline{\tau})$. The distance between $\overline{j}$ and the closer of these two points is $\eta + \overline{\tau}$. The difference between the slopes of $\overline{L}_1$ and $\overline{L}_2$ is 2, so the intersection of these lines lies outside $\overline{S}$ provided that $\eta + \overline{\tau} > 2\overline{\tau}$. But this follows from the basic facts.

Now consider the parameter p/q. We use the same notation. This time we have $|m(L_1)| > \omega - 4\kappa$. By Lemma 25.6 we have $|m(L_2)| < 4\kappa$. In order for the two lines L_1 and L_2 to intersect in $\Upsilon(\overline{S})$, at least one of these lines must have an endpoint in the segment J mentioned above. But for the super tricky index j, the point $(0, j)$ does not lie in J. Hence L_2 intersects J. But then $|m(L_2)| = 1$. By Lemma 26.2, the lines L_2 and $\overline{L}_2$ have the same Y-intercept. The same argument as for the parameter $\overline{p}/\overline{q}$ now shows that the light points on L_1 lie outside $\Upsilon(\overline{S})$. $\square$

This completes the proof of the Copy Lemma. The proof of the Copy Lemma completes the proof of Theorem 0.7.

References

[**DeB**] N. E. J. De Bruijn, *Algebraic theory of Penrose's nonperiodic tilings*, Nederl. Akad. Wentensch. Proc. **84**:39–66 (1981).

[**DF**] D. Dolyopyat and B. Fayad, *Unbounded orbits for semicircular outer billiards*, Annales Henri Poincaré **10** (2009) pp 357-375

[**G**] D. Genin, *Regular and Chaotic Dynamics of Outer Billiards*, Pennsylvania State University Ph.D. thesis, State College (2005).

[**GS**] E. Gutkin and N. Simanyi, *Dual polygonal billiard and necklace dynamics*, Comm. Math. Phys. **143**:431–450 (1991).

[**H**] W. Hooper, *Renormalization of polygon exchange transformations arising from corner percolation*, Invent. Math. **191.2** (2013) pp 255-320.

[**Ko**] S. Kolodziej, *The antibilliard outside a polygon*, Bull. Pol. Acad Sci. Math. **37**:163–168 (1994).

[**M**] J. Moser, *Is the solar system stable?*, Math. Intelligencer **1**:65–71 (1978).

[**N**] B. H. Neumann, *Sharing ham and eggs*, Summary of a Manchester Mathematics Colloquium, Jan 25, 1959, published in Iota, the Manchester University Mathematics Students' Journal.

[**S1**] R. E. Schwartz, *Outer Billiards on Kites*, Annals of Math Studies **171**, Princeton University Press (2009).

[**S2**] R. E. Schwartz, *Outer billiards and the pinwheel map*, J. Mod. Dyn. **2**:255-283 (2011).

[**S3**] R. E. Schwartz, *Outer Billiards, Polytope Exchange Transformations, and Quarter Turn Compositions*, preprint (2013).

[**T1**] S. Tabachnikov, *Geometry and billiards*, Student Mathematical Library 30, Amer. Math. Soc. (2005).

[**T2**] S. Tabachnikov, *Billiards*, Société Mathématique de France, "Panoramas et Syntheses" 1, (1995).

[**VS**] F. Vivaldi and A. Shaidenko, *Global stability of a class of discontinuous dual billiards*, Comm. Math. Phys. **110**:625–640 (1987).

Index